PROCEEDINGS OF THE DYNAMIC MEASUREMENT AND CONTROL FOR AUTONOMOUS MANUFACTURING WORKSHOP

Edited by:
Tsai Hong, Roger Eastman, Roger Bostelman, Hui-Min Huang
National Institute of Standards and Technology (NIST)

Brian McMorris
SICK, Inc.

Dates:
October 10 - 11, 2007

Location:
Loyola College of Maryland
Columbia, Maryland

Sponsors:
NIST, Loyola College in Maryland, and SICK, Inc.

NISTIR 7575

Proceedings of the Dynamic Measurement and Control for Autonomous Manufacturing Workshop

Edited by:
Tsai Hong
Roger Eastman
Roger Bostelman
Hui-Min Huang
U.S DEPARTMENT OF COMMERCE
Technology Administration
National Institute of Standards and Technology
Intelligent Systems Division
Gaithersburg, MD 20899-8230

and

Brian McNorris
SICK, Inc

May 2009

U.S. DEPARTMENT OF COMMERCE
Gary Locke, Secretary
NATIONAL INSTITUTE OF STANDARDS AND TECHNOLOGY
Patrick Gallagher, Acting Director

TABLE OF CONTENTS

DISCLAIMER 2
POLICY 2
EXECUTIVE SUMMARY 3
1. BACKGROUND 5
2. WORKSHOP INFORMATION 6
   2.1 Participating Organizations 6
   2.2 Workshop Agenda 7
3. WORKSHOP RESULTS 10
   3.1 Day 1 Breakout Sessions Outcome 10
      A. *Requirements for dynamic 6DOF metrology for automated general assembly* 10
      B. *Control and perception needs for automated guided vehicles (AGVs)* 13
      C. *Information needs for real-time process monitoring and control* 15
      D. *Automated manipulators, their subcomponents and controls needed for dynamic manufacturing* 17
   3.2 Day 2 Wrap-up Discussion and Results 20
      A. Needs identified during the workshop: 20
      B. Action Items and Associated Volunteers 20
4. INITIAL ACTION ITEMS RESULTS 22
   4.1 Formulating New Projects for Identified Critical Needs 22
   4.2 Further Investigation on Industrial Needs 22
   4.3 Identification of Current Standards 23
      4.3.1. Terminology and system characteristics 23
      4.3.2. Safety and performance 24
      4.3.3. 3D metrology 25
5. REFERENCES 27
APPENDIX: WORKSHOP PRESENTATIONS 28

DISCLAIMER

Certain trade names and company products are mentioned in the text or identified in certain illustrations. When permissions for publication were provided to National Institute of Standards and Technology (NIST), the presentations of the participating companies or products are included in the Appendix to facilitate communications among the participants. In no case does such an identification imply recommendation or endorsement by the NIST, nor does it imply that the products are necessarily the best available for the purpose.

The opinions expressed in this Workshop Report are those of the workshop participants' and are not the official opinions of NIST.

POLICY

It is NIST's policy to use the International System of Units (SI). However, some of the units used in the workshop presentations and papers are in U.S. customary units because of the intended audience. Conversions from the U.S. customary units to SI have been made where possible.

EXECUTIVE SUMMARY

The Dynamic Measurement and Control for Autonomous Manufacturing workshop was held on October 10 - 11, 2007 at the Columbia Graduate Center of Loyola College in Maryland. Forty-eight people attended the one and one-half day event which was moderated by Roger Bostelman and Tsai Hong of the National Institute of Standards and Technology (NIST) with assistance from Roger Eastman of Loyola and cosponsored by Brian McMorris of SICK, Inc. Attendees included current and potential users of autonomous manufacturing equipment, manufacturers of general assembly equipment, robotic arms and automated guided vehicles, and of machine vision and three-dimensional (3D) sensors, system integrators, government representatives from NIST and National Science Foundation (NSF), and academics with research specialties in computer vision and robotics.

The workshop was a part of an effort by NIST Manufacturing Engineering Laboratory (MEL) to assist industry in articulating the general requirements for advanced automation in manufacturing. The stated goal of the workshop was:

> "To collect community input on requirements for the operation of next generation manufacturing robots, automated guided vehicles (AGVs), and intelligent assist devices in dynamic, changing environments. Specific topics to be considered are perception needs for dynamic visual servoing in autonomous assembly and requirements for the safe and effective operation of robot arms and AGVs in dynamic environments."

Current robot and AGV installations are generally rigid, sensor-poor and expensive to reconfigure. Standards, metrics, and measures are required to assist general assembly manufacturers in introducing new technologies and deploying flexible, sensor-rich systems suitable for dynamic operating environments. These new systems could enable equipment re-use, fewer dedicated installations, and faster and more flexible plant reconfiguration, resulting in significant cost savings and higher productivity. Industry, academia, and government research institute input and collaboration will enable NIST to articulate requirements and to explore methodologies to assist the manufacturers in evaluating and validating new technologies.

The first morning of the workshop consisted of a series of presentations given by manufacturers highlighting their needs, by academics on the state of current research in relevant areas, and by sensor vendors on the relevant capabilities. Also presented was the status of ASTM E57 3D Imaging System standards development efforts. In the afternoon, the attendees participated in four breakout sessions on particular challenges before reassembling to share the results. The focuses of the break out groups are

- A. dynamic metrology and perception on the assembly line,
- B. needs for enhanced robot control systems to operate in dynamic environments,
- C. the requirements for advanced AGV use in dynamic environments, and
- D. real-time sensing to validate and update virtual simulations.

Most of these topics originally arose in the Fall 2006 Smart Assembly Workshop held at NIST. Workshop discussions continued the next morning in a plenary setting to summarize and to develop action items. The results are summarized below:

1. Perception systems for automation in dynamic environments must be comprehensive, pervasive and redundant. In scenarios such as a robot grasping a moving part off an assembly line, a single, narrowly focused sensor will not be sufficient. A single sensor may fail, may not be robust enough for the task, may not sense other objects that could become obstacles, may not be able to adequately sense humans in the workspace to prevent accidents, may have the wrong wavelength or modality to be useful in a task, or may have a fixed resolution while the task requires sensing at multiple resolutions. Perception for such scenarios will require sensor fusion and control logic to facilitate arbitration between multiple subsystems. Sensory modeling must be improved so that the performance can be compared and evaluated.

2. While most participants agreed on the general nature and need for next generation robots and perception systems, the group wanted to see more specifically defined and challenging scenarios to focus future discussions and to direct future research, much as the DARPA Grand Challenge does for mobile robotics. This issue was established as a follow-up action item for this group. Related to this issue, common themes in the breakout session discussions were the need for terminology standards, high level robot control vocabularies; common interfaces that support operations in dynamic environments, and articulation of requirements to identify standard useful tasks.

3. The need for performance standards and measurement techniques for localization was echoed by multiple breakout session groups. Dynamic metrology is required to provide reference measurements so that the performance of the perception systems can be evaluated for safety, for AGV navigation, and for moving part manipulation in changing environments. To judge whether an AGV or a robot arm can perform a task which involves motion with respect to a second moving object, we need techniques for calibrating and measuring absolute and relative motions.

4. Moving to next generation robotics with safe and reliable performance in dynamic environments will require the development of an entire cycle of commercial interests that can produce the components to be integrated into next generation assembly lines. As such, system integrators and standard specifications are needed for the purposes of component interfacing and test and evaluation. This concept and technology, in its entirety, is young and evolving, therefore, may take time to mature. Also hindering the progress is a chicken-and-egg problem with new technology, because few companies can afford to invest in research or the application of a technology until the technology has been demonstrated as successful. To alleviate the dilemma, participants proposed to use testbeds and high fidelity simulation systems so that the technology can be validated before commercialization should begin.

5. Software is an important element of the perception systems for the control the motion of robots, AGVs and other automation devices. Capabilities are required to validate the software to ensure safety and to achieve reliable performance.

1. BACKGROUND

In most manufacturing assembly and material transport environments, parts are delivered on moving lines to be picked up and attached to the base assemblies, which are usually also moving. Currently, an automated facility must be equipped with expensive, custom designed mechanisms, such as specific bowl feeders and conveyances and fixtures to control the motion and the positioning of the parts. Systems that are customized for certain current needs are typically not scalable enough to handle next generation manufacturing that features unstructured and dynamic environment. Such a new environment requires systems that are real-time controllable, agile, adaptable, flexible, and reconfigurable. Also required are next-generation safety technologies that advance human-robot collaboration to a new level.

New technologies, including advanced perception and advanced planning and motion control are required to achieve this next generation manufacturing capability. It is key to have sensor technology that can perceive the position of a part under unconstrained motion and either inspect the part or direct a robot to manipulate it while still in motion. The technology must also be able to perceive possibly co-existent humans and other moving objects and generate corresponding safety actions. To enable this functionality, a method is needed to continuously measure the six-degree of freedom (6DOF) location and orientation of an unconstrained moving object. Existing affordable pose measurement systems are too inaccurate, brittle and slow. No standards exist to evaluate the accuracy of such systems or to guide users in their adoption. There are a number of candidate non-contact technologies with good potential including stereo cameras, laser triangulation, structured light, interferometry, scanning ladar, flash ladar and monocular geometric matching smart cameras, but there is a lack of common terminology and common vendor accuracy measurements leading to confusion in user comparisons and marketplace hurdles. As a result, few are in use and many manufacturing tasks are not automated or make use of humans and assistive devices. A reference standard for dynamic 6DOF pose measurement would advance the technology by establishing metrics for the evaluation of these systems and techniques. In addition to automated assembly tasks on a traditional line, advanced perception systems would be useful in material transport and other industrial tasks associated with manufacturing.

Terminology standardization is recognized as a means of facilitating next generation dynamic manufacturing. Given the size, rich history, and ongoing research and development efforts of manufacturing industry, there is abundant vocabulary that is either existent or evolving. For example, practitioners begin to use such terms as Next Generation Robots, but with different meanings. Some use the term to mean robots with the capabilities of higher payloads or inherent safety design. Standardization can be beneficial. In addition, there are terms that are developed in the defense and homeland security unmanned systems (UMS) community that can be explored for the manufacturing automation purposes as the industry is moving more and more toward autonomous operations and intelligent manufacturing for the sake of safety and productivity. The Terminology part of the NIST Autonomy Levels for Unmanned Systems (ALFUS) Framework [B.h.1] presents these opportunities that should be worthy of exploration.

The ability to measure the positions and orientations of components as they move would result in considerable cost savings in applications such as automobile manufacturing by replacing expensive fixed installations with more intelligent combinations of sensing and automation. The ability would also enable greater flexibility and adaptability for U.S. manufacturers and better enable them to compete with foreign firms where greater investments have been made recently in robotic technology. There would be substantial, immediate benefits in industry segments like automotive and airplane assembly, but the technology is fundamental and could be widely applied. A reference standard would also assist the academic community in establishing clear performance metrics for research systems and algorithms.

2. WORKSHOP INFORMATION

2.1 Participating Organizations

Sponsoring Organizations
    National Institute of Standards and Technology
    Loyola College in Maryland
    SICK, Inc.

End Users – Presentation Topics
    General Motors – Autonomous Assembly
    Washington Post - Newspaper Manufacturer
    US Postal Service - Distribution
    General Dynamics Electric Boat – Shipbuilding
    SICK, Inc. - Next Generation Robots
    Ford Motor Company – Process Modeling

Academics and Government – Presentation Topics
    Carnegie Mellon University – AGV Control
    Purdue University - Dynamic Visual Servoing
    SRI International – Visual Simultaneous Localization and Mapping (VSLAM)
    NIST Building and Fire Research Laboratory - 3D Imaging Systems Standards

Sensor, Robot and AGV Manufacturers – Current and New Products
    <u>Sensors:</u>
        SICK, Inc. – Non-Contact Measurement
        Automated Precision, Inc. – 6 Degrees of Freedom (DOF) Measurement
        TYZX – Stereo Imaging
        Shafi (USA) – Stereo Imaging
        Mesa Imaging – Range Camera

    <u>Robot Arms:</u>
        Barrett – WAM Robot Arm
FANUC   Robotics
        Vecna – Bear Robot

Automated Guided Vehicles:
    Egemin Automation – Automated Guided Vehicles
    FMC – Automated Guided Vehicles

Demonstrations of Products
    Mesa Imaging – Range Camera
    API - 6D measurement system
    SICK - laser measurement
    TYZX – 3D Stereo Vision Platform
    Barret - robot arm

2.2 Workshop Agenda

Wednesday, October 10, 2007

8:00-8:20 AM   Welcoming Remarks (5 min. each) – Room 230/210

    Opening Remarks: Roger Eastman, Professor
    Loyola College in Maryland

    NIST, Manufacturing Engineering Laboratory (MEL), Intelligent Systems Division
    Overview: Al Wavering, Acting Deputy Director, MEL

    Overview: Roger Bostelman, Manager
    Intelligent Control of Mobility Systems Program

8:20 – 10:00 AM   End Users (15 min. each + 5 min. Q/A, set-up)
    What are the main, prioritized manufacturing issues?

    End User 1 – Automobile Manufacturer
        Roland Menassa, General Motors – Autonomous Assembly
    End User 2 – Newspaper Manufacturer
        Conrad Rehill, Washington Post
    End User 3 – Distribution
        Joyce Guthrie, USPS
    End User 4 – Shipbuilding
        Ken Fast, GDEB (presented by Roger Bostelman, NIST)
    End User/Facilitator 5 – Next Generation Robots
        Brian McMorris, SICK, Inc.

10:00 - 10:20 AM   Q&A / Discussion

10:20 – 10:30 AM   Break

10:30 – 11:10 AM	Academia and Government (10 min. each) Past / present research

                    Carnegie Mellon University, George Kantor – AGV Control
Purdue   University, Avi Kak - Dynamic Visual Servoing
                  SRI International, Moti Agrawal - VSLAM
                  NIST Building and Fire Research Laboratory, Alan Lytle - Standards

11:10 – 11:30 AM	Q&A / Discussion
11:30 – 12:15 PM	Sensors, Robots and AGV's - (3-5 min. each) Current/New Products to Support Dynamic Measurement and Control for Autonomous Manufacturing

Sensors:
  SICK,   Brian McMorris
               Automated Precision, Inc., Tom McLean
  TYZX,   Gaile G. Gordon
  Shaf   i (USA), Adil Shafi
               Mesa Imaging – Range Camera, Peter Hunt
               Robot Arms: Barrett, FANUC Robotics and Vecna, Claude Dinsmoor

               Automated Guided Vehicles: Egemin Automation and FMC

12:15 - 1:45 PM	Lunch and Exhibits – Room 230/210 Demos of products: – Rooms 208, 251 - Mesa Imaging – Range Camera - API - 6D measurement system - SICK - laser measurement - TYZX – 3D Stereo Vision Platform - Barret - robot arm
1:45 AM – 1:55 PM	Charge to Breakouts Groups Roger Eastman, Breakout information
2:00 – 3:45 PM	Breakouts

Breakout Groups to Deliberate and Draft Research Recommendations and Performance Metrics Requirements

A.   Requirements for dynamic 6DOF metrology for automated general assembly
        Moderators: Roger Eastman, Loyola University and Tsai Hong, NIST
  Room   : 262

B.   Control and perception needs for automated guided vehicles
        Moderators: Roger Bostelman and Stephen Balakirsky, NIST
  Room   : 270

C. Perception needs for real-time process monitoring and control
Moderators: Mike Shneier and Hui-Min Huang, NIST
Room: 280

D. Robot arms, their subcomponents and controls needed for dynamic manufacturing
Moderators: Fred Proctor and John Horst, NIST
Room: 272

3:45 – 4:00 PM   Break

4:00 – 5:00 PM   Plenary Session – Room 230/210
Breakout groups to present summaries

Thursday, October 11, 2007

8:15 - 8:30 AM	Welcoming Remarks Roger Bostelman, NIST
8:30 - 9:00 AM	End User 6 – Automobile Manufacturer Dimitar Filev, Ford – Process modeling
9:00 – 10:30 AM	End User Panel Discussion: Where To Go From Here: Standards and Technology Roadmap
10:30 – 10:45 AM	Break
10:45 – 11:45 AM	Academic Panel Discussion: Where To Go From Here: Research Roadmap
11:45 – 12:00 AM	Wrap-up Summary Tsai Hong, NIST Where to go from here: Report and other follow on activities

12:00 PM   Adjourn

3. WORKSHOP RESULTS

3.1 Day 1 Breakout Sessions Outcome

The workshop organizers identified the following critical issues to be addressed by the participants:

A. dynamic metrology and perception on the assembly line,
B. needs for enhanced robot control systems to operation in dynamic environments,
C. the requirements for advanced AGV use in dynamic environments, and
D. real-time sensing to validate and update virtual simulations.

Four breakout groups were formed correspondingly. Each was assigned a central theme that highlighted a key issue in the envisioned dynamic manufacturing environment. Also provided were a corresponding vision and a preliminary information statement intended to foster discussions. Participants were assigned to the groups according to their affiliations. The objective is to have vendors, users, and developers evenly distributed in the breakout sessions to facilitate well-rounded viewpoints of the issues. Note that the actual participations were different from the assigned as some participants felt that they had contributions to offer for the other groups and attended multiple sessions.

The group activities and results were described in the following sections.

A. *Requirements for dynamic 6DOF metrology for automated general assembly*

Moderators: Roger Eastman, Loyola College in Maryland and Tsai Hong, NIST

Group Assignments:

Name	Organization
Jane Shi	General Motors (GM)
Jonathan St. Clair	Boeing
Peyush Jain	Goddard Space Flight Center
Steve Freedman	SICK
Peter Kamp	SICK Germany
Kam Lau	Automated Precision, Inc
Zaifeng Chen	Automated Precision, Inc
Dave Strzegowski	General Dynamics Robotic Systems (GDRS)
Jamie Nichol	Vecna Technologies
Roger Eastman	Loyola Univ.
Avi Kak	Purdue University
Tsai Hong	NIST
Daniel Dementhon	National Science Foundation

Information Provided Prior to the Workshop

Vision

To achieve flexible and reconfigurable automation of manufacturing processes through sensor technology that can perceive the poses of a part in motion by dynamic 6DOF pose measurement.

Preliminary information to foster discussion:

In many manufacturing assembly environments, parts are delivered on moving lines and must be picked up and attached to the corresponding part being manufactured, which is also moving. Current technology typically requires the line to stop while an action is performed or a measurement taken. To achieve flexible and reconfigurable automation of assembly processes, it would help to have sensor technology that can dynamically perceive the pose of a part, in 6DOF. What issues may be involved in using improved or advanced sensor technologies to achieve accurate and robust 6DOF measurements? Are current sensor technologies up to the task? Where are technological advances needed? In new sensors? In the improvement of current sensors through factors such as advanced resolution and frame rate? In improved algorithms for motion and pose analysis? Will the solution require the fusion of data from multiple sensors?

Results of 6 DOF Metrology Group Discussions

During their discussion, the workshop participants listed the following elements as important to the development and success of the envisioned new technology.

Roadblocks and Challenges

These elements were determined to be problems faced by the new technology:
- Advanced 6DOF perception is a young technology, in its initial phases and with maturity perhaps 10 years away, and will face a question of economic viability.
- An entire commercial ecosystem will be required, involving sensor manufacturers, system integrations, robots manufacturers, standards organizations, and others.
- A successful system will need to exhibit continuous adaptation to changing conditions, as such, the system will have to be very complex, hence very difficult and expensive to develop.
- A successful system will need to be robust and have clearly specified capabilities and limitations. The extremely large numbers of possible parts that may be needed and situations that may occur in the envisioned dynamic and unstructured manufacturing environments will make this new technology difficult to achieve.

Concrete Scenarios

The participants put together a few cases of interest to manufacturers for the phased development and evaluation of 6DOF sensors. The scenarios are listed in order of difficulty:
1. First case: mating of two rigid parts under dynamic conditions, as typically encountered in automotive general assembly.
2. Second case: mating of a flexible part, such as a hose, to a moving rigid part.

3. Third case: mating of parts, rigid or not, that would involve complex path planning, such as an attachment inside a vehicle or base assembly.
4. Fourth case: manipulation of non-rigid attachments associated with parts, such as an electronic automotive part with a number of wiring leads attached.

Solutions

The group listed the following characteristics as important to a successful advanced perception solution:
- A successful 6DOF perception system should be comprehensive, pervasive, redundant, multi-level and multi-resolution. Such a system can perceive an entire scene, sense the position of a part over a wide range of distances, and be robust against the isolated failure of components or sensors.
- A successful sensor system that is faster and more accurate might be a substitute for more intelligence, as better information about the world can reduce the requirements for reasoning.
- A successful perception system should be a part of an overall solution, balanced with other concerns. For example, the need for more accurate sensing during a robotic pick-up operation may be mitigated by improving the compliance of the grippers, allowing less precise sensing.

What types of sensors?

The group produced an initial listing of the categories of sensors that are likely to be used in a solution, shown below:
- Single camera
- Multiple cameras (stereo)
- Range sensors
- Structured light
- Laser scanning
- Flash ladar
- Force/torque sensors

Research needs

The group considered areas in which new research will be required to solve the problem. The following are its findings:
- Research should be conducted in the use of multiple, heterogeneous sensors, including fusing sensor data, enabling cooperation among sensors, and arbitrating between sensors with conflicting data.
- Research should be conducted in the modeling of sensor performance in static and dynamic environments. Research should also be conducted in the subsequent estimation methods of the sensory measurement confidence levels.

Sensor Requirements/Metrics

While the number of possible manufacturing applications is very large and hard to easily characterize, the group made some initial progress in the area of spatial and temporal tolerances, as described below:
- For general assembly, the position of an object should be measured to about 0.32cm

(1/8 in).
- For specialized assembly tasks, these tolerances may be tightened to 1/4000cm (1/1000 in).
- The latency of the sensor system should be minimal, ideally near 0. All things equal, a higher sampling rate is better.
- The performance of a sensor system should be commensurate with the motion statistics of the manufacturing environment, taking into account range, velocity and accelerations.

Standards Needs
- The pose of a part needs to be measured with a specified standard deviation with respect to each of the 6 degrees of freedom when the line is moving at a given speed with specified statistical properties.

B. *Control and perception needs for automated guided vehicles (AGVs)*

Moderators: Roger Bostelman and Stephen Balakirsky, NIST

Group Assignments:

Name	Organization
James Wells	GM
Joyce Guthrie	USPS
Brian McMorris	SICK
Peter Hunt	Mesa Imaging
Joe Stanford	Automated Precision, Inc
Gaile G. Gordon	TYZX
William T. Townsend	Barrett Technology
Mark Longacre	FMC
Brad Byle	Egemin Automation
Andreas Hofmann	Vecna Technologies
Moti Agrawal	SRI International
George Kantor	CMU
Roger Bostelman	NIST
Steve Balakirsky	NIST

Information Provided Prior to the Workshop

Vision
To achieve improved AGVs that can be more rapidly deployed and can operate in dynamic, unstructured environments

Preliminary information to foster discussion:
In their current form, AGVs are useful but limited on the tasks that they can perform. They are not able to access all unstructured areas of a plant, may require particular plant floor design to accommodate them, and may need special fixtures for loading and unloading. The problems become more difficult when people, parts and a mixture of AGVs are moving in the environment. What issues must be solved before the technology

is universal, flexible and robust enough to support dynamic manufacturing? What are the issues in drive systems, in absolute and dynamic positioning systems, and in AGV–coordination among themselves, with central systems, and with humans? What are the issues in modeling and simulation, such as simulating the simultaneous motions of a fleet of AGVs to achieve high factory efficiencies? Are new, radical designs needed to perform required tasks, such as motion in spaces design primarily for humans or in carrying manipulators for tasks like installing a wiper blade on a moving automotive assembly line?

Results of 6 DOF Metrology Group Discussions

During their discussion, the workshop participants listed the following elements as important to the development and success of the technology:

Roadblocks
The following elements were determined to be problems and questions faced by the new technology:
- A useful AGV must be capable of localization, finding its location relative to a map well. Questions related to localization were determined to be:
  o What is the best use of an internal GPS for absolute positioning and mapping?
  o What is the process to find a vehicle location with respect to a known point (localization)?
  o What tolerance is needed for the dynamic measurements? Would an accuracy or repeatability of 10 mm be adequate?
  o Will localization be achieved with multiple sensors or one "magic" sensor?
  o Is there a need for retrofitting current facilities with localization capabilities to enable them to handle unstructured environments? For example, is there a need to convert from a rigid set of markers/tracks, installed on the floor or at other places, to a flexible AGV system with dynamic localization capabilities? If so, what would be the cost?
- Advanced AGVs will need to have functionality that meets user demand and costs that are affordable. Questions are:
  o What's keeping users from using more AGV's, AGC's (smart-carts), and forklifts/tuggers? Is there a chicken-and-egg problem, where apparent lack of demand is holding back development?
  o What are the infrastructural costs to supporting an intelligent AGV?
  o Is there a demand for robot arms on vehicles?
  o What would be the advanced safety standards for mobile robots? There is a lack of clear definitions for safe operation of an AGV.
  o Can AGVs be made taller to have a higher work volume, yet are still stable?
  o Can they be made faster, yet still safe and have adequate stopping distances?
  o Can they be made easy to use and flexible, so they can be quickly brought across assembly lines?
  o Can they be made scalable, from low to high volume, and manual to autonomous in operation?
- Other technical issues in AGV development and acceptance:
  o From the perspectives of real time control and scheduling, should the control be

integrated on the vehicle or distributed in the workplace?
   - Are there low cost sensors that are robust enough, come with adequate support, yet meet the new needs in safety and performance?
   - Is battery technology holding back AGV performance?
   - Are there adequate standards for AGVs and military vehicles, along the lines of TRL (Technology Readiness Level) or ALFUS, for commercial use?

Required solutions and research needs
- We need to develop better standards, performance metrics, and system specification methodology related to:
  - Safety standards for AGVs to categorize and quantify risks to humans.
  - Collaborative AGVs that work together on tasks.
  - Capabilities of arms on vehicles that enable mobile manipulation.
  - Specifications of plans, standard task vocabularies and levels of autonomy (ALFUS) to enable easier descriptions of vehicle capabilities.
  - Criteria for ease and intuitiveness of AGV programming, including
    - Better high-level programming languages,
    - Improved user interfaces for direct teaching modes.
  - Performance under varying environmental conditions, such as lighting.
  - Localization and mapping, including
    - How accurate must a vehicle be to safely navigate and tow objects, parts trays, and carts?
    - How accurate is a map produced by sensors?
    - How accurately can an AGV localize itself?
- We need lower cost and better vehicular technology for:
  - Brake locks that can fully support both emergency and protective stops.
  - Suspension and intelligent compensation technology that can accommodate variations in suspension. For example, soft suspension can lead to localization problems related to dead reckoning.
  - Mobile manipulation with integrated arm(s) on vehicle.
  - Coordinated control and planning for single or multiple AGVs.
  - Sensors for functionality and safety.

C. Information needs for real-time process monitoring and control

Moderators: Mike Shneier and Hui-Min Huang, NIST

Group Assignments:

Name	**Organization**
Conrad Rehill	Washington Post
Jacqueline LeMoigne-Stewart	Goddard Space Flight Center
Stephan Schmitz	SICK Germany
Clarence Burns	Automated Precision, Inc
Jay Li	Automated Precision, Inc
Adil Shafi	Shafi (USA)
Eric Beaudoin	GDRS

James Albus	NIST Fellow
Jeremy Zoss	Southwest Research Institute
Rama Chellappa	University of Maryland
German Londono	Purdue University
Mike Shneier	NIST
Hui-Min Huang	NIST

Information Provided Prior to the Workshop

<u>Vision</u>

To achieve manufacturing line efficiency and quality improvement by better acquisition and use of on-line (perception-based) and a priori (model-based) information.

<u>Preliminary information to foster discussion:</u>

Knowledge of the current state of the assembly line can be critical in achieving efficiency, in avoiding bottlenecks and quality problems, in planning for higher efficiencies or line redesign, and in keeping virtual line simulation coordinated with the real-world. Advanced sensors may be able to play a role in monitoring the assembly line and performing dynamic metrology on parts in motion, advancing the current field of machine vision to better adapt to unstructured environments. What information do users need from perception systems to manage and control their process? What does the virtual assembly information contribute to the process? What are the dynamic metrology capabilities in perception systems that improve process control in unstructured environments?

Results of 6 DOF Metrology Group Discussions

<u>Roadblocks</u>
- The vision statement was found to be in line with the industrial problems and no adjustments were made.

<u>Problems:</u>
- The labor force has a shortage of workers skilled in automated machine operation.
- It is difficult to integrate new technology (machine vision, etc.) into existing systems.
- Intelligent equipment is expensive and hard to evaluate.
- Knowledge acquisition and representation are difficult problems for automated manufacturing.
- It is difficult for automated systems to detect and identify critical events from current information sources.
- Current safety requirements can hinder human operations in robotic environments – there is a need to clearly identify what happens and take actions to mitigate the effect. For example, having humans nearby should not necessarily shut the robots down.
- Small batch jobs and customized orders exemplify the problem of the variety of things to be measured. These point to the need for automation.
- Equipment can malfunction in dirty environments. Examples include simple situations like dust on sensors and uneven floors that interfere with AGV dead reckoning.
- There are many sources of errors and exceptions. For example, error models can

become ineffective.
- It is important, but difficult to, have post-operation verification—make sure that the operation succeeds what is supposed to be done, especially in safety related issues.
- With the contributions of a representative of the newspaper industry, the discussion yielded a number of fruitful items specific to that industry. They are:
    o Inserts of newspapers count as a significant portion of a newspaper company's revenue. An important requirement is to deliver them to where the advertisers want, according to zip codes, streets, or other demographic concerns.
    o Heavy time constraints exist to identify particular pallets for inserts, followed by their loading and delivery.
    o There is a need to label particular bundles for accurate tracking. Currently need to produce 700,000 copies daily.
    o Vision systems may be a good technology to identify and retain knowledge about the bundles.
    o Bundles may break in the process and mess up the counting system.
    o Current accuracy rate is about 98.6% for the Washington Post AGV operation.

Solutions
- Need to understand and model the full operation of AGVs. All the possible exceptions must be listed and programmed into the systems to become parts of the model.
- Need metrics to evaluate the robustness/costs/performance of the implemented algorithms.

Research needs
(Did not get to this topic.)

Requirements/Metrics
(Did not get to this topic.)

Standards
(Did not get to this topic.)

D. *Automated manipulators, their subcomponents and controls needed for dynamic manufacturing*

Moderators: Fred Proctor and John Horst, NIST

Group Assignments:

Name	Organization
Maravas, Michael	USPS
Theodore Bugtong	Goddard Space Flight Center
Wolfgang Bay	SICK Germany
Yuanqun Liu	Automated Precision, Inc
Tom McLean	Automated Precision, Inc
Mark Bankard	GDRS

Brian Zenowich	Barrett Technology
Claude Dinsmoor	GE Fanuc
Daniel Theobald	Vecna Technologies
Johnny Park	Purdue
John Horst	NIST
Fred Proctor	NIST

Information Provided Prior to the Workshop

Vision
- To achieve better flexibility and ease of deployment and operation of automated manipulators in dynamic environments
- Customers: domain experts, not technology experts, need assistance in technology choices.
- Deployment benefits accrue mostly for one-of-a-kind installations; operation benefits accrue for everyone.
- Dynamic: everything can change: environment, process, product

Preliminary information to foster discussion:
To achieve better flexibility and ease of deployment, robot arms, intelligent assist devices and other programmable devices for automated handling of material will need to operate more and more in dynamic environments where people, parts, conveyers and AGVs are moving. In the advanced case, robot arms will be mounted on AGVs or humanoid platforms and operating in a dynamic environment where they will be interacting with people. What issues are there in advancing the technology so this can be accomplished safely and efficiently? Are there issues in arm design, in programmability, or in standardization of interface between robot controller and other system elements such as PLCs and sensors? Are there issues in underlying control theory or calibration and validation of system performance? What elements of the systems will need to be enhanced to take advantage of advanced 3D sensors?

Results of 6 DOF Metrology Group Discussions

Roadblocks
The group voted to prioritize the identified roadblocks. Participants could vote for multiple items. The size of a vote (following each item) represents a measure of group consensus towards prioritization. The ones without a vote were seen as either lower priority or not common issues across the entire industry.
- Achieving safe and flexible robotics and enabling collaboration with humans--resulting in new applications: 14
- Improving software-based safety chain, including possibly revisions to the existent standards. This step would enable the achievement of the other roadblocks on this list: 7
- Making AGVs with robot arms easy to use and to set up, including deployment and development: 8
- Enabling conformance tests, performance tests, and verification against a specification: 8

- Eliminating incompatibility, e.g., issues with data exchange, connectors, programming languages: 5
- Dealing with obsolescence of AGVs.
- Hardening systems for work in dirty environments.

The highest-ranking requirements are further elaborated as below:

R1: Achieving safe and flexible robotics: enabling collaboration
- o Technologies needed: better collision sensing and avoidance.
- o Metrics needed: what is damaging to a person? For example, the head injury criteria from auto industry might be applicable.
- o Standards needed: software-based safety chain.
- o Research needed: what constitutes safe behaviors (varies widely across applications and robot types); what are effective ways for robots and humans to collaborate?

R2: Ease of use, setup, including deployment and development
- o Technologies needed: application development techniques that domain experts are familiar with.
- o Standards needed: interface to robotic functions that support operations in a dynamic environment.

R3: Conformance tests, performance tests, verification against a specification
- o Metrics needed: performance for the identified range of human-robot interactions; trust or perception of safety.
- o Standards needed (other than the safety standard): motion detection, frame rates, reaction bandwidth, or any that supplements the current Robotic Industries Association (RIA) standards.
- o Research needed: trust and perception of safety, risk assessment.

3.2 Day 2 Wrap-up Discussion and Results

Participants engaged in a discussion that summarized the first day's findings and looked to follow-up actions. The following are the results:

A. Needs identified during the workshop:
   a. Better and more complete standards in the areas of:
      - Interfaces for equipment and software interconnection
      - Terminology –lists of common tasks and commonly used terms
      - Evaluation metrics and methods for sensor, AGV, robot and overall system performance

      The development efforts should be synchronized with the defense industry activities.
   b. More use of scenarios and competitions to drive development:
      - Industry associations could pool resources to establish challenges.
      - Establish a list of the challenges and competitions for researchers to study.
      - Simulation challenges are much easier to be set up. NIST runs one on a simulated AGV.
   c. Manufacturers need to provide more information on automation issues and impacts:
      - Form consortia to collect general issues and observations on challenges faced.
      - Establish individual research on specific industry needs.
   d. Better supporting technologies:
      - Establish the capability to seamlessly run tests from virtual to real.
      - Develop techniques to monitor the automation progress based on requirements and metrics.

B. Action Items and Associated Volunteers
   a. Overarching goal: develop standards for interfaces, performance, metrics, and terminology, e.g., robot safety and software components.
   b. Become involved in RIA standard processes:
      - Jim Wells, Roger Bostelman, and Brian McMorris will begin looking into how to start process of developing future standards. The types of standards lab tests that can be beneficial to the industry include: feasibility, conformance, and performance.
      - Hui-Min Huang will look into terminology standards: identify manufacturing terms, look for existing standards.
   c. Create well-defined challenge scenarios:
      - Joyce Guthrie, Jane Shi, Roger Bostelman, and Stephen Balakirsky.
      - Non-rigid components, e.g., carpets, cable harnesses, hoses are to be included.
   d. Develop cooperation with industry associations and systems integrators to target research on "needs:"

- Automotive Industry Action Group (AIAG), Mechanical Contractors Association of America (MCAA), Material Handling Industry of America (MHIA), RIA, and Society of Manufacturing Engineers (SME) are among the relevant associations.
- Jim Wells has experience in this and should be a lead.
- Names of systems integrators can be forwarded to Jim Wells and Roger Bostelman.

e. There seems to be no trade organizations that focus on systems integrators.
f. Set up Workshop mail group:
   - Fred Proctor
g. Complete workshop report:
   - Workshop organizers
h. Next workshops/meetings and other interesting forums –
   - Combine need-focused meetings with major conferences.
   - Robot Industry Forum, Orlando, Nov. 2007
   - IEEE Conference on Computer Vision and Pattern Recognition (CVPR, which tends to be academic), International Manufacturing Technology Show (IMTS, September 2010), International Robots, Vision & Motion Control Show hosted by RIA (June 2009) [2, 3, 4].

4. INITIAL ACTION ITEMS RESULTS

As of this report date, work has already begun on the identified action items and the results are described in the following sections.

4.1 Formulating New Projects for Identified Critical Needs

To address a key sensory requirement for dynamic manufacturing, NIST has embarked on a dynamic 6DOF pose measurement project. The goal is to devise a method to continuously measure the locations and orientations of an unconstrained moving object. Current automated assembly systems typically measure the pose of an object only in highly constrained situations, such as parts moving at a fixed speed in a rigid conveyance, or by stopping the assembly line to sense the precise position of the part. Locating and tracking an arbitrary object under unconstrained motion is very difficult as majors issues exist for most of the sensing technologies. For example, optical camera-related technology may involve loss of 3D information through projection. The 6DOF related technology must require the pose to be reconstructed from the data in real-time for which the equations and algorithms are not yet fully understood. A reference standard for dynamic 6DOF pose measurement would advance the technology by establishing metrics for the evaluation of these systems and techniques.

The project will develop methods for continuous measurement for manufacturing applications such as automobile manufacturing and for evaluating the sources of error in the measurements, including finding out how to minimize the errors. This will require techniques to calibrate the reference and test systems, to synchronize measurements for comparisons, to evaluate the raw sensor data that is used to compute 6DOF pose, and to track the contribution and propagation of errors in subsystems.

The ability to measure the positions and orientations of components as they move would result in considerable cost savings in applications such as automobile manufacturing. The ability will allow expensive fixed installations to be replaced with more intelligent combinations of sensing and automation, and, thus, better enable US manufacturers to compete with foreign firms where greater investments have been made recently in robotic technology. Although our current focus is on the substantial, immediate benefits in industry segments like automotive and airplane assembly, the technology is fundamental and could be widely applied. A reference standard would also assist the academic community in establishing clear performance metrics for research systems and algorithms.

Roger Bostelman and Stephen Balakirsky began coordinating space allocation with the NIST machine shop and designing a testbed for a NIST exploratory project on vehicle navigation through unstructured environments. The testbed will include a robot arm mounted on overhead rail to enable a number of automation scenarios. This testbed will serve as a beginning for well-defined scenario testing as they arise.

4.2 Further Investigation on Industrial Needs

This workshop and its proposed future standards development effort were discussed in the 15th Annual Robotics Industry Forum that Roger Bostelman, Brian McMorris, and Jim Wells attended in Orlando on November 7-9, 2007.

Jim Wells discussed with RIA participants at the Orlando Robotics Industry Forum about research on "needs."

4.3 Identification of Current Standards

Hui-Min Huang took an action item of researching the current robotic standards in the areas of vocabulary for tasks and systems. The results include a collection of ISO, ANSI, and RIA standards that mostly deal with low level devices, coordinate systems, geometry/kinematics, programming, and limited performance evaluation. They are summarized later in this section. This finding points to a possible broad-scope structure for robotic standards that may encompass multiple levels of abstraction for the knowledge. Aspects of task structures, a general purpose unmanned systems terminology, and ontology may all be covered in the structure. It would be interesting to find out how the concept of autonomy levels can be applied in the manufacturing domain. Corresponding terms like Unmanned Flexible Manufacturing System (UFMS), and Unmanned Workstation (UWS) might be explored to embed various levels of operator interactions.

4.3.1. Terminology and system characteristics

The following standards are identified and are listed according to the publishing organizations.

ISO

ISO 14539: 2000

Manipulating industrial robots -- Object handling with grasp-type grippers -- Vocabulary and presentation of characteristics
Categories: types of handling, grasps, coordinate systems and sensing in object handling, types of grasp-type grippers, types of end effectors, elements of grasp-type grippers, types of grasp-type grippers, types of fingers, finger control, clamping elements, robot interfaces, safety in grasps and grasping

ISO 9787: 1999

Manipulating industrial robots -- Coordinate systems and motion nomenclatures
Content: world, base, mechanical interface, and tool coordinate systems, robot motion, robot axes

ISO 9946: 1999

Manipulating industrial robots -- Presentation of characteristics
Manufacturer shall provide: application, power source, mechanical structure, working space, coordinate system, external dimension and mass, base mounting surface, mechanical interface (how end effectors are mounted on robotic wrists), control, task programming and program loading, environment, velocity, resolution, performance criteria, safety

ISO 11593: 1996

Manipulating industrial robots -- Automatic end effector exchange systems -- Vocabulary

and presentation of characteristics
Categories: external shape, main dimension, position and orientation in coupling procedure, coupling and releasing forces, load, magazine interface, tool exchange time

ISO 8373: 1994/1996
Manipulating industrial robots - Vocabulary
An amendment ISO 8373:1994/Amd 1:1996 and a corrigendum ISO 8373/Cor.1:1996 followed.
Categories: general terms, mechanical structure, geometry and kinematics, programming and control, performance

AIAA:
R-103: 2004
AIAA Recommended Practice: Terminology for Unmanned Aerial Vehicles and Remotely Operated Aircraft

AIAA S-066
Standard Vocabulary for Space Automation and Robotics (1995)

ASTM:
E 2521 – 07a
Standard Terminology for Urban Search and Rescue Robotic Operation

E 2544
Standard Terminology for Three-Dimensional (3D) Imaging Systems

IEEE:
IEEE 100-2000
The Authoritative Dictionary of IEEE Standards Terms (Seventh Edition, 2000)

NIST:
NIST SP 1011-I-2.0
Autonomy Levels for Unmanned Systems Framework, Volume I: Terminology, Version 2.0

4.3.2. Safety and performance

ISO:
ISO 10218-1, -2: 2006
Robots for industrial environments — Safety requirements

ISO 9409-1, -2, -3: 2004
Manipulating industrial robots — Mechanical interfaces

ISO 9506-1, -2: 2003
Industrial automation systems — Manufacturing Message Specification

ISO 9283: 1998
Manipulating industrial robots - Performance criteria and related test methods

ANSI/RIA
ANSI/RIA RI5.05-1: 1990
Point-to-Point and Static Performance Characteristics - Evaluation

ANSI/RIA RI5.05-2: 1992
Path-Related and Dynamic Performance Characteristics - Evaluation

ANSI/RIA RI5.05-3: 1992
Reliability Acceptance Testing - Guidelines

NIST:
NIST SP 1011-I-2.0
Autonomy Levels for Unmanned Systems Framework, Volume II: Framework Models, Version 1.0

4.3.3. 3D metrology

Existing standards for 3D metrology, useful in defining terminology, artifacts and protocols that might be relevant to this effort exist [6]. In addition, there are ongoing efforts for developing performance standards for imaging systems. Below list three of these efforts:

• ASTM: 3D imaging sensors [7]
The ASTM Committee E57 on 3D Imaging Systems has been investigating standards for 3D imaging sensors, with the Building and Fire Research Lab (BRFL) at NIST taking a leadership role. The BRFL conducted workshops with sensor vendors and other interested parties in 2003, 2005 and 2006, and has done work to define terminology and initial protocols. The current focus is on static, large-scale metrology.

• IACMM: Non-contact metrology
The International Association of Coordinate Measurement Machine Manufacturers (IACMM) is supporting work on Optical Sensor Interface Standard (OSIS). This standard is intended to aid the integration of non-contact sensor technologies into traditional contact coordinate measurement machines. The standard has three elements on physical interfaces, software interfaces, and calibration. The latter effort covers accuracy specification and validation for the 3D data from non-contact sensors. The scope of this effort may cover 3D imaging systems of interest and dynamic and 6DOF performance is not emphasized. NIST participates in this project.

• EMVA: Machine vision sensor performance
The European Machine Vision Association (EMVA) has the 1288 standard effort, "Standard for Measurement and Presentation of Specifications for Machine Vision Sensors and Cameras." The scope of the standard currently covers monochrome digital area scan cameras and should be extended to line and color cameras. The format could be a model for

reporting 3D imaging performance. The standard is developed in a modular fashion, with each module defining a physical sensor model for characterizing sensor response, a protocol for testing the characteristics, and a format for presenting and analyzing the results.

5. REFERENCES

1. *Autonomy Levels for Unmanned Systems (ALFUS) Framework, Volume II: Framework Models Version 1.0,* NIST Special Publication 1011-II-1.0, Huang, H., Ed., National Institute of Standards and Technology, Gaithersburg, MD, December 2007

2. http://www.cvpr.org/

3. http://www.imts.com/

4. http://www.robotics.org/

5. Hong, T.S., et al., *Dynamic metrology: Evaluation of 3D Imaging Systems for Dynamic Sensing Applications in Manufacturing*, NIST Draft Report, January 2009

6. *Proceedings of the LADAR Calibration Facility Workshop June 12 – 13, 2003,* NIST Internal Report, NISTIR 7054, Cheok, G., Ed., October 2003

7. http://www.astm.org/COMMIT/COMMITTEE/E57.htm

APPENDIX: WORKSHOP PRESENTATIONS

Overview: Roger Bostelman, Manager
    Intelligent Control of Mobility Systems Program 29

End Users
    Automobile Manufacturer
        Jane Shi, et al., General Motors – Autonomous Assembly 34
    Newspaper Manufacturer
        Conrad Rehill, Washington Post 41
    Distribution
        Joyce Guthrie, USPS 48
    Shipbuilding
        Ken Fast, GDEB – presented by Roger Bostelman, NIST 51
    Next Generation Robots
        Brian McMorris, SICK, Inc... 57

Academics and Government
    Carnegie Mellon University, George Kantor – AGV Control 68
    Purdue University, Avi Kak - Dynamic Visual Servoing 75
    SRI International, Moti Agrawal – VSLAM ......…............................ 88
    NIST Building and Fire Research Laboratory, Alan Lytle – Standards 97

Sensors, Robots and AGV's
    SICK, Brian McMorris 111
    TYZX, Gaile G. Gordon 134
    Shafi (USA), Adil Shafi 140
    API, Kam Lau 146

Dynamic Measurement and Control for Autonomous Manufacturing Workshop

Roger Bostelman, Tsai Hong, Stephen Balakirsky,
Elena Messina, Hui-Min Huang, *NIST*
Roger Eastman, *Loyola College in Maryland*
Brian McMorris & Steve Freedman, *SICK*

Loyola College in Maryland, Columbia, MD
October 10-11, 2007

workshop slide #1

Vision

- "Industrial robots guided by machine vision have the potential to revolutionize manufacturing processes, improving repeatability, cycle rate, reliability and safety on the plant floor, while reducing costs associated with labour and fixturing."
    - Vision-guided robotics - March/April 2007 - Examining the technology's impact on the plant floor
    - By Mary Del Ciancio
- "with vision guidance, robots can be deployed increasingly in places where robotic automation was not imaginable before, and you can see that this has a powerful effect on the landscape of manufacturing and the way we will lay out our plants of the future."
    - Babak Habibi, President, Braintech

workshop slide #2

Roadblocks

- lighting – ambient, sunlight, low light, part appearance
- training
- integration
- physical constraints of the sensor system
- location/logistics - where the system will go and who will operate it.
- financial issues - system cost, ROI
- practical issues - upkeep, maintenance and training
- control of the environment – airborne particles, abrasives, vibrations
- timing – e.g., latency control between the robot controller
- data type – 2D, 3D
- communications - with robots, single or multiple sensors
- fixtureless/moving parts
- ….

Industrial Robot Roadblocks

- 95%+ of industrial robots are used without sensors in the outer loop

Robot Challenges
- Systems Integrations – largest portion of robot supply chain
    - Largely disconnected from robot providers
    - There are few established standards for system design
    - Few tools available for comprehensive modeling
- Need programming and I/O support

Henrik I Christensen   KUKA Chair of Robotics, Georgia Tech
hic@cc.gatech.edu

Workshop Challenge

- How do we address these roadblocks?
    - Research
    - Technology
    - Standards
    - Performance Metrics

Current

- **automotive industry** - assembly and processing of engine and body components.
- **food industry** - pick products from conveyors for packaging into individual containers or cartons.
- **pharmaceutical industry** - locate medical supplies on moving belts for packing into shipping cartons.
- **metalworking industries** - finding metal castings on pallets and loading CNC machines to make finished component products
- **bin-picking applications** today that just a few years ago were thought to be impossible." Roney:
- "**Material handling is the low-hanging fruit** that this technology can be used to capitalize on today,"
    - **Typical [2D] applications** [are] picking from a stationary or moving conveyor, pallet loading/unloading, conveyor tracking, and component assembly
    - **Typical [3D] applications** are auto-racking and bin-picking. We are also seeing interest for robotic deburring and material removal applications (e.g., find and deburr parts) McLauglin, Boatner

WELCOME!
Attending Organizations

- National Institute of Standards and Technology
- Loyola College of Maryland
- SICK - US and Germany
- General Motors
- Ford
- Washington Post
- US Postal Service
- Goddard Space Flight Center
- Mesa Imaging
- Automatic Precision, Inc.
- TYZX
- Shafi (USA)
- General Dynamics Robotic Systems
- Barrett Technology
- GE Fanuc
- FMC
- Egemin Automation
- Vecna Technologies
- Southwest Research Institute
- Purdue University
- Stanford Research Institute
- Univ. of MD
- CMU
- National Science Foundation

workshop slide #7

Final Agenda
Wednesday, October 10, 2007

7:30-8:00 AM — Continental Breakfast
8:00-8:20 AM — Welcoming Remarks
  Opening Remarks: Roger Eastman, Professor, Loyola University
  NIST, Manufacturing Engineering Laboratory (MEL), ISD Overview: Al Wavering, Acting Deputy Director, MEL
  Overview: Roger Bostelman, Manager, Intelligent Control of Mobility Systems Program
8:20 – 10:00 AM — End Users *(15 min. each + 5 min. Q/A, set-up)*
  What are the main, prioritized manufacturing issues?
10:00 - 10:20 AM — Q&A / Discussion
10:20 – 10:30 AM — Break
10:30 – 11:10 AM — Academics and Government (10 min. each)
  Past / present research
11:10 – 11:30 AM — Q&A / Discussion
11:30 – 12:15 PM — Sensors, Robots and AGV's - (3-5 min. each)
  Current/New Products to Support Dynamic Measurement and Control for Autonomous Manufacturing
12:15 - 1:45 PM — Lunch and Exhibits
  Demos of products:
1:45 AM – 1:55 PM — Charge to Breakouts Groups
2:00 – 3:45 PM — Breakouts
  Breakout Groups to Deliberate and Draft Research Recommendations and Performance Metrics Requirements
3:45 – 4:00 PM — Break
4:00 – 5:00 PM — Plenary Session
  Breakout groups to present summaries
5:00 – 6:00 PM — Cocktail Hour
6:00 – 9:00 PM — Dinner *(On your own.)*

workshop slide #8

Final Agenda
Thursday, October 11, 2007

7:30 - 8:15 AM	**Continental Breakfast**
8:15 - 8:30 AM	**Welcoming Remarks**
	Roger Bostelman, NIST
8:30 - 9:00 AM	**End User 6 – Automobile Manufacturer**
	Dimitar Filev, Ford – Process modeling
9:00 – 10:30 AM	**End User Panel Discussion:**
	Where To Go From Here: Standards and Technology Roadmap
10:30 – 10:45 AM	**Break**
10:45 – 11:45 AM	**Academic Panel Discussion:**
	Where To Go From Here: Research Roadmap
11:45 – 12:00 AM	**Wrap-up Summary**
	Tsai Hong, NIST
	Where to go from here: Report and other follow on activities
12:00 PM Adjourn	

workshop slide #9

Where To Go From Here?

- Combine and Prioritize across breakouts:
  - Research
  - Technology
  - Standards
  - Performance Metrics

workshop slide #10

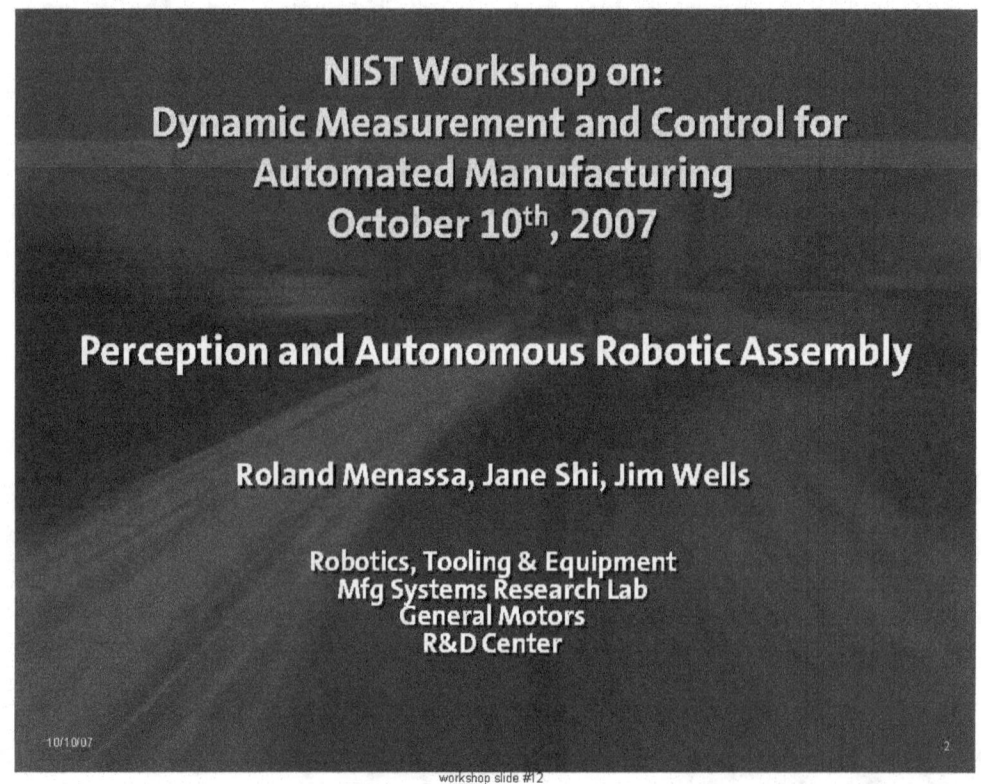

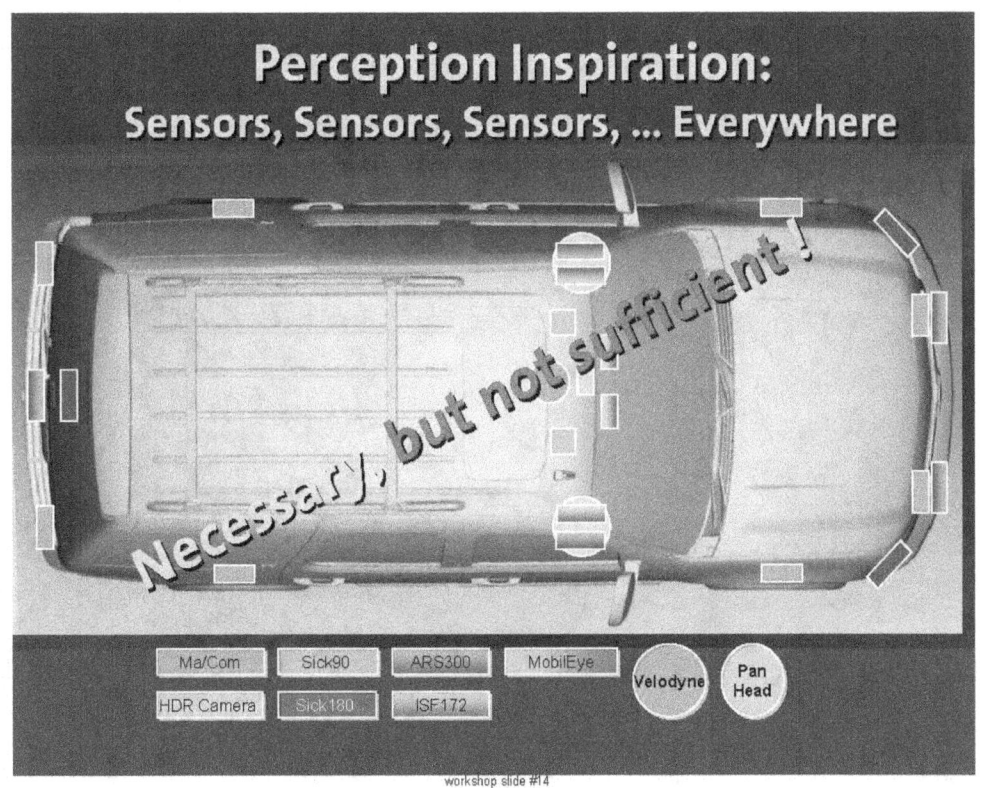

Perception Inspiration: Sensors Serve their Purposes

On-road driving

Obstacle avoidance

Parking maneuvers

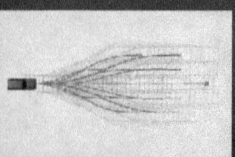
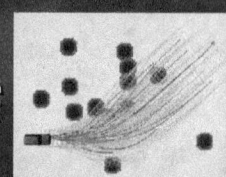
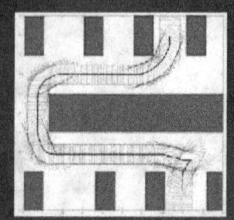

Courtesy of CMU Tartan Racing

Autonomous System of High Capabilities

Requires
Intelligent High Level Planning,

Robust and Adaptive Low Level Behavior Control

With Adequate Dynamic System Response

One Example: NIST's 4D/RCS

workshop slide #17

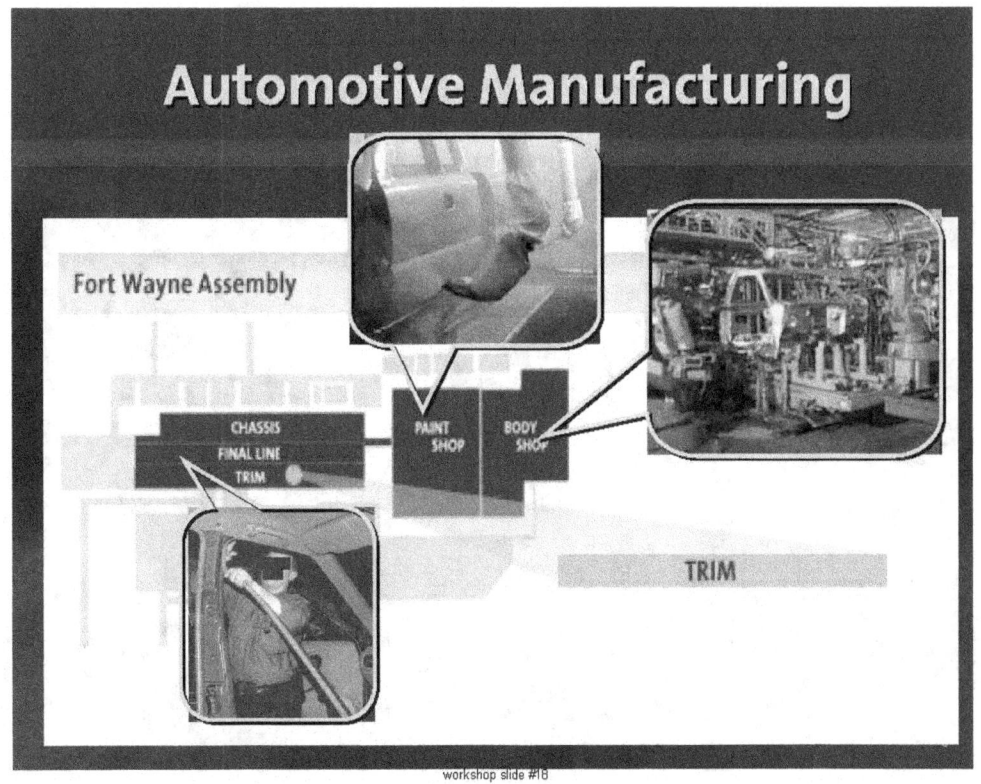

workshop slide #18

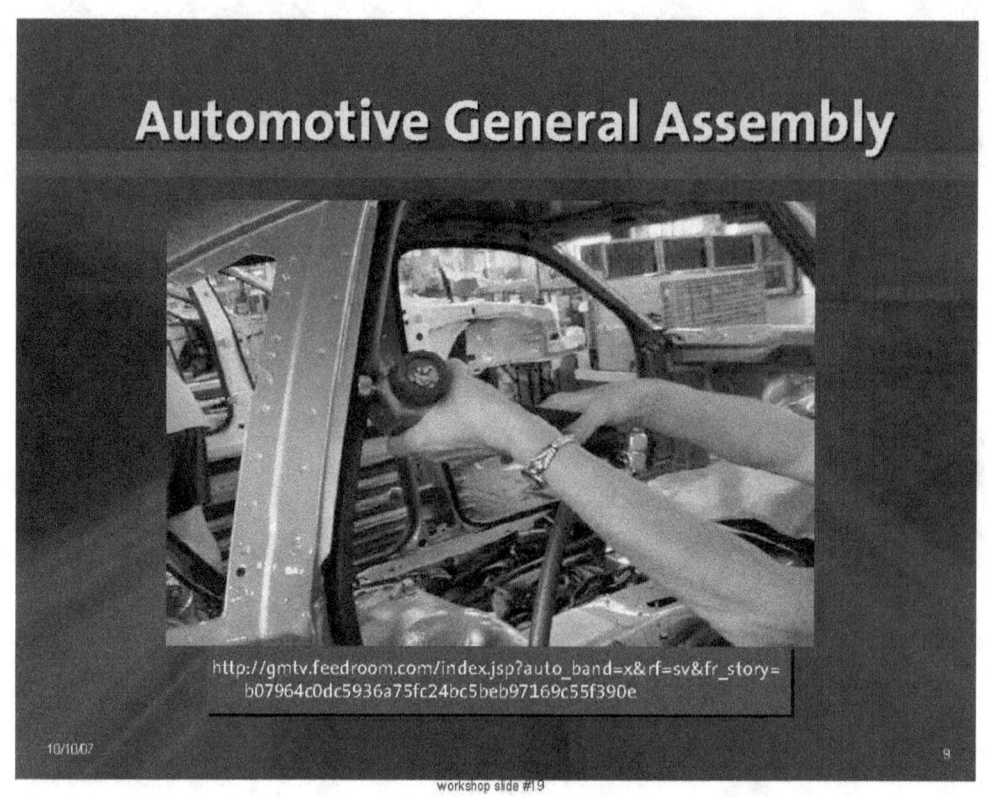
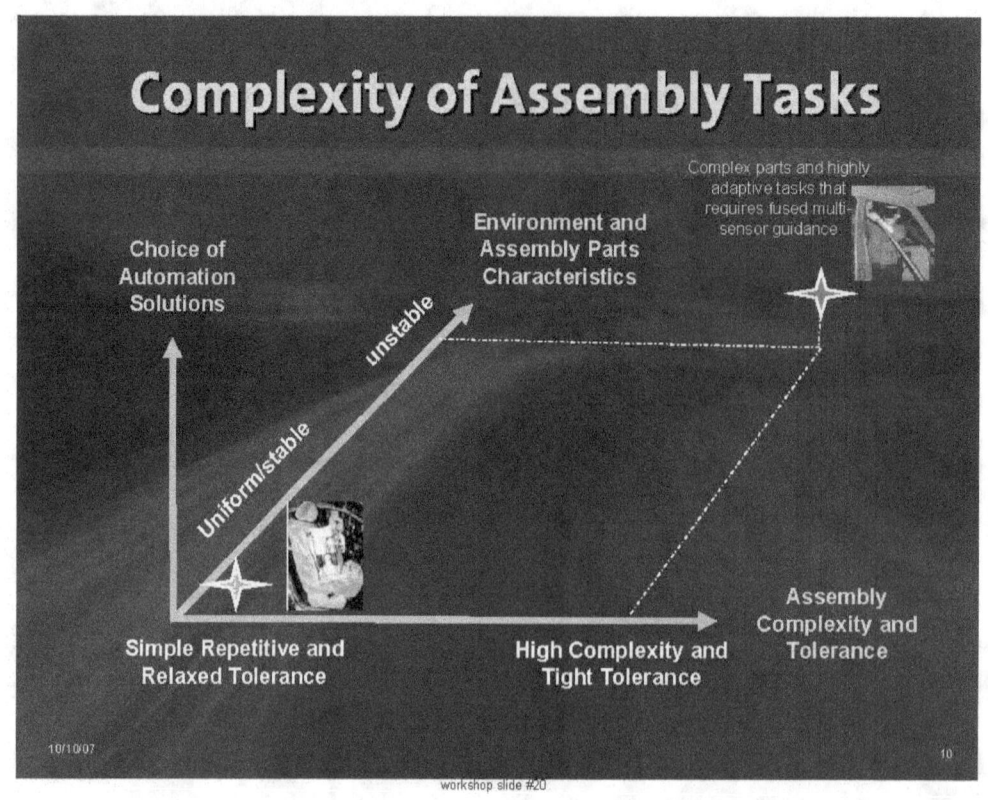

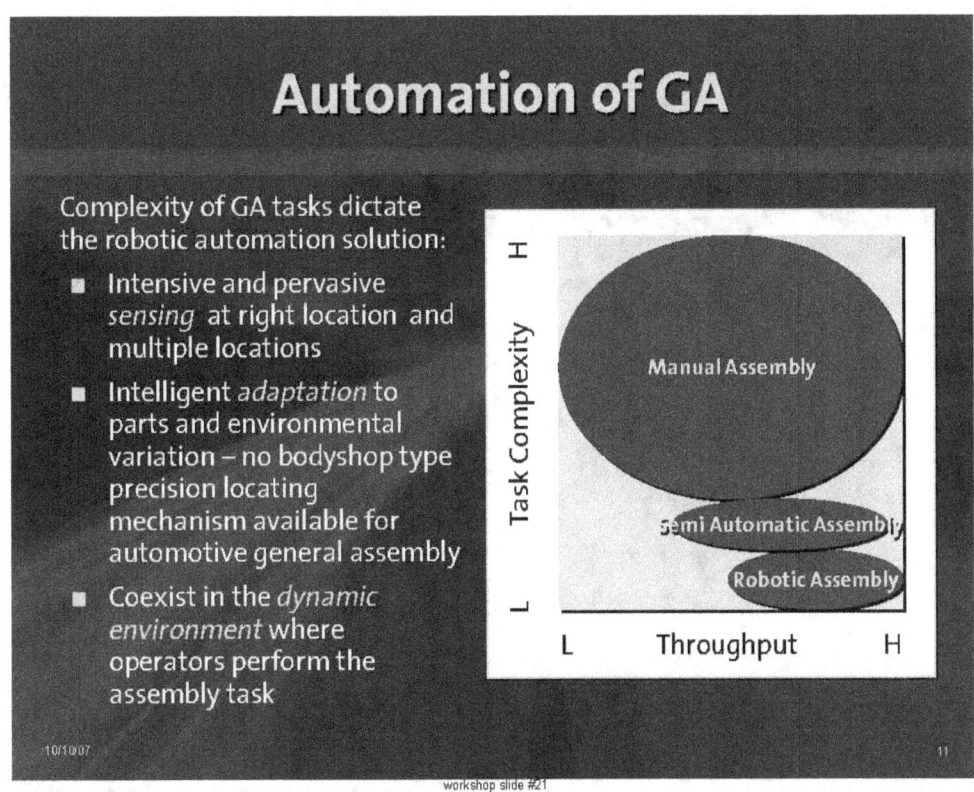

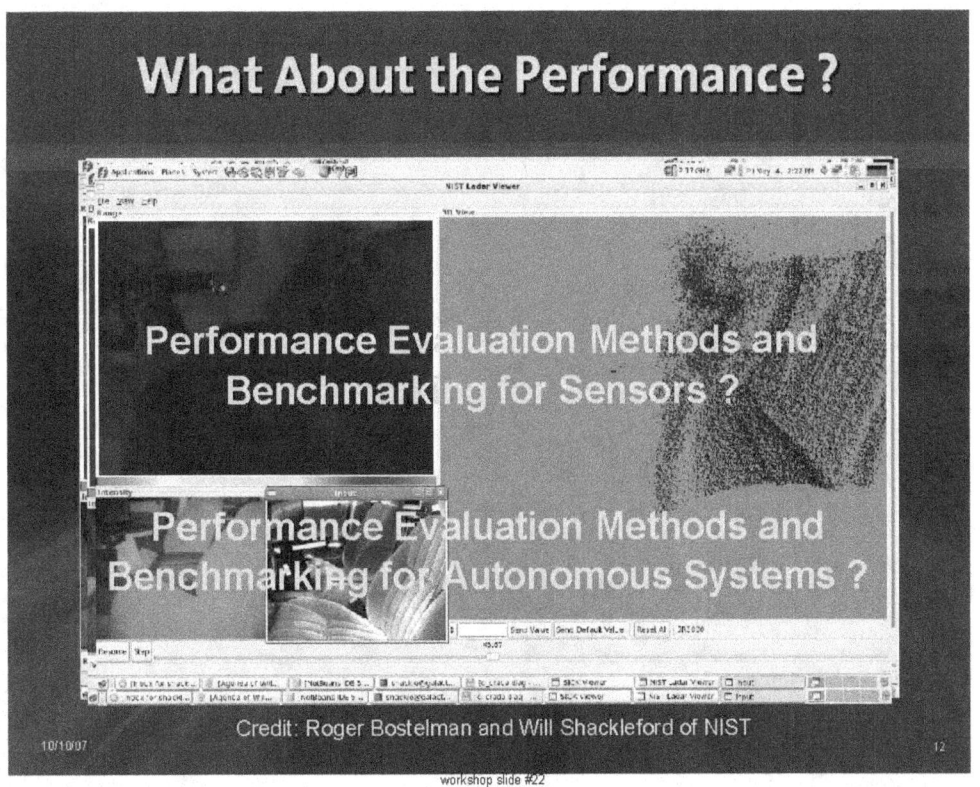

The Washington Post

Automation Challenges

Conrad Rehill

Manager of Systems Engineering

Production Facilities

- NW Washington, DC
  - Newsroom & Business Functions
- Springfield, Virginia
  - Printing: DC, Virginia, National Edition
  - Weekday pre-print advertising insert
  - Sunday pre-print advertising collation
- College Park, Maryland
  - Printing: Maryland & North

Changing Business

- The product: News & Advertising
- Decline in newspaper readership nationwide
  - Internet, Cable Television … Recycling (?)
- Increasing selectivity on advertisers part
  - increased focus on targeted delivery of advertising with smaller zones / higher penetration

Newsprint Delivery

- FMC PICS system (installed 1998)
  - Redundant control system
  - 16 vehicles, 14 required to meet peak demand
  - 150,000 kg Virginia / 100,000 kg Maryland per night

What affects reliability ?

- System
  - controls
  - vehicles

- Human factors
  - Operators
  - Maintenance

- Environment
  - vibration / shock
  - RF environment

Vibration & Shock

Vibration & Shock

- Original condition of floor never met AGV vendor specifications
  - flatness, incline, slab gap
- Continual replacement of vibration-worn vehicle electronics and suspension components
- The fix: cut new epoxy resin "lanes" in floor providing a seamless travel path

Analyzing the Fix

- New travel lanes required shot blasting 1" deep channels in existing floor
- Poor cleanup of shot material becomes embedded in vehicle wheels
- Shot-embedded wheels wear 3/4" grooves in areas of floor not yet repaired with epoxy resin

Analyzing the Fix

- Excessive vibration while traversing worn areas of floor and shock of transitioning from worn to unworn levels increase vehicle breakdown rate
- Approach critical number of vehicles simultaneously down (14 of 16 required to meet daily peak demand)

Continuous Monitoring

- Use PC/104 board with standard WiFi
  - onboard 12-bit A-D conversion
  - single-axis accelerometer
  - tap into vehicle onboard power
  - RS-232 serial data tap to eavesdrop on host-vehicle communication: location
  - dump data across WiFi at vehicle closest approach to data repository

Other Opportunities

- Vision systems: color registration
  - Replacement of existing system costly
  - Maintenance of existing system questionable
  - Use GNU tools with COTS PCI-based counter, DIO, and frame grabber cards
- Vision systems: bundle tracking
  - need to have knowledge of bundle identity prior to application of a label

Color Registration

- The task: identify the registration pattern within a field that may contain confusing data

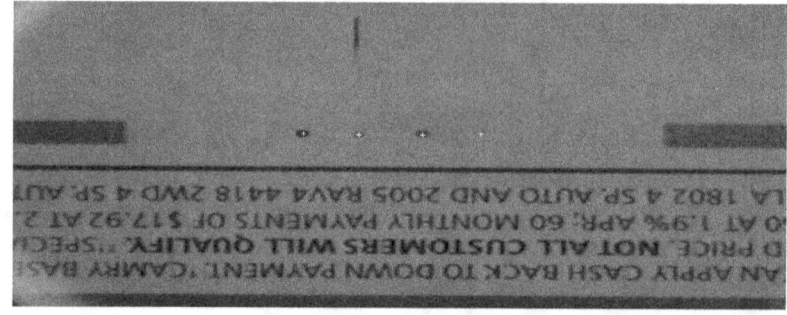

THANK YOU !

Questions ?

workshop slide #37

Sensors On The Robotic Containerization System
(RCS)

Engineering
Package and Material Handling Development

Joyce Guthrie

RCS

- **USPS has purchased and installed 167 systems**
  - Each system contains 2 gantry style robots
  - USPS has decided to add 10 more RCS III systems to the fleet, bringing the total to 177 systems by 2009
- **First RCS I system was installed in Santa Ana, CA in 2000**
- **Last RCS II system was installed in St. Louis, MO in September 2007**
- **USPS is the largest user of Gantry robots in the world**

RCS

- Sensors
    - We use a wide variety of sensing with the robotic system
        - Collision sensing on the robot arm and EOAT (end of arm tool)
            - This is a Applied Robotics sensor that is attached to the arm prior to the EOAT for detection of ay collisions in the sideways position
        - Search sensor
            - To detect if a shelf in the container is there
        - Detection sensor
            - Pallet/container detection (to detect what is in the position)
            - Docking station for present and type
            - SMM detection stand

RCS

- ❏ Sensors (continued)

  - Safety sensing
    - Levels of access control sensing
      - 1st level is physical door
      - Door interlocks
      - Light curtains
      - On gantry pop up hard stops
      - Switches to determine which zone the robot is in
      - Pull cords
      - Plexiglas doors with interlock switch at pick-up station

  - Tray present sensing for zone control

RCS

- ❏ Sensors – Lesson Learned
  - ➢ Issues
    - Dirt effects them
    - Sensitivity range (if adjustable)
    - Type of material on the retro-reflective (bounce back issue)
    - Alignment sensitivity
    - Obsolescence
    - Connector configurations

Shipbuilding: Automation Issues

**Ken Fast,
General Dynamic Electric Boat**
supplied as generic information

860-433-6432, kfast@ebmail.**gdeb**.com

Presented by:
Roger Bostelman, NIST
at the
Dynamic Measurement and Control for Autonomous Manufacturing Workshop

Oct. 10, 2007

Overview/outline

Shipbuilding offers some unique challenges in manufacturing:
- large, heavy structures
- long build time
- single-item build
- limited repeat
- (limited indoor fabrication space)
- large range of operations/disciplines

Everything is custom build due to long build time.

EUROP Sectorial Report on Industrial Robot Automation

Naval shipbuilding has some additional requirements

- specialty materials - high strength steels, stainless, nickel/copper
- relatively tight tolerances

Probably costs more as a result

Measurement tasks

- piece/part verification
- flat cut plate – CNC cut now
- rolled/shaped plate
- assembly layout - not much fixturing
- assembly verification
- equipment mating holes/surfaces – in place!
- equipment/assembly installation, alignment
- large unit join alignment
  – E.g., join two 40' x 40' x 100' weighing 100 tons

Unique Measurements
(at least for submarines)

- circularity
- reference planes
  - arbitrary references that may not be on the part
    - E.g., 4' away from the part

Automation – state of industry

- all piece parts design-to-cut (CAM)
- mostly automated pipe bending
- some robotic welding
- some automated sheet metal cutting/forming
  - waterjet, laser, oxy-fuel
- some automated plate forming

(non) Automation

- permanent fixturing/workcells
  - custom build, single item limits use of
- dedicated floor layout from unit to unit
  - long build schedule, limited repeat, and large structures cause difficulty in maintaining
- automated material movement
  - large, heavy items, outdoor transit, and changing layout limit use of
- automated material movement
  - most material delivery requires lifting (sometimes heavy), safety issues limit
- automated welding/coatings application
  - odd shapes, constrained spaces, material specifications limit use of
  - (multi-pass welding, thick sections, pre-heating)

Design product model issues

- limited manufacturing information in design
  - limited repeat
    - E.g., a lot of odd shapes (subs)
  - 3rd party design
    - may not have exactly what's needed on the drawing to make the part
- reference planes, odd shapes

Existing measurement techniques

- part layout marking (laser etch)
- optical surveys
- photogrammetry
- laser alignment
- laser tracking
- "string" lines
  - still works well!

Current work efforts

- part "families" - similar tasks/shapes with dedicated, but flexible, workstations
- (more) dedicated floor layout
- flexible, accurate location technologies - iGPS
- increased use of layout marking
- increased integration of manufacturing data with design product model
- robotic/automated welding
- robotic/automated coatings

Wish List

- large area reference line/plane projection
  - (a-la laser-level used in construction) – although, not visible in space, can't always measure to it from a tape measure
- multiple planes
- relative alignment (not level)
- solid steel interference
- line-of-sight issues
- automated welder
- field deployed, easy setup, easy program, multi-pass

Thanks Ken!

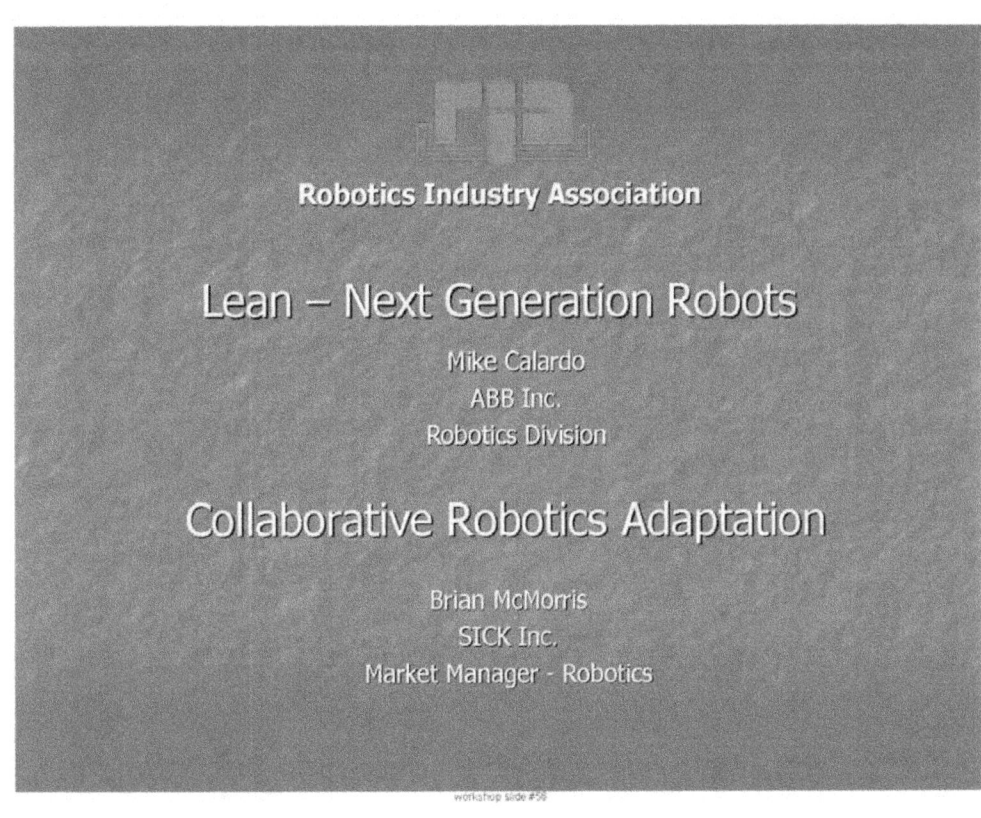

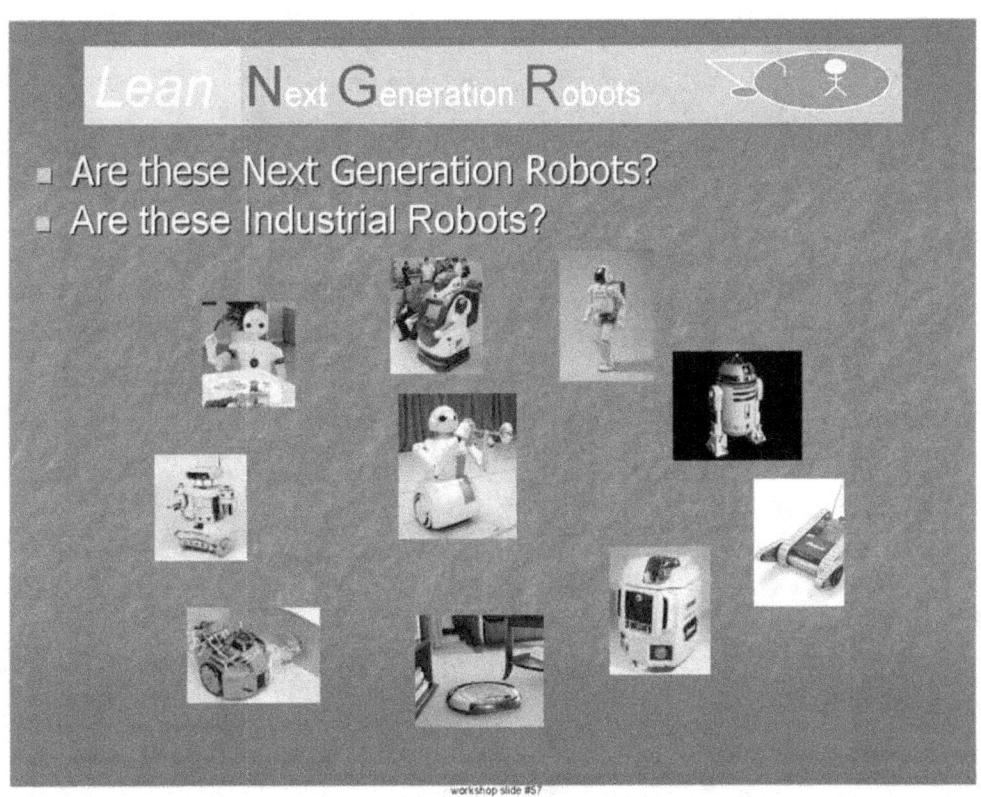

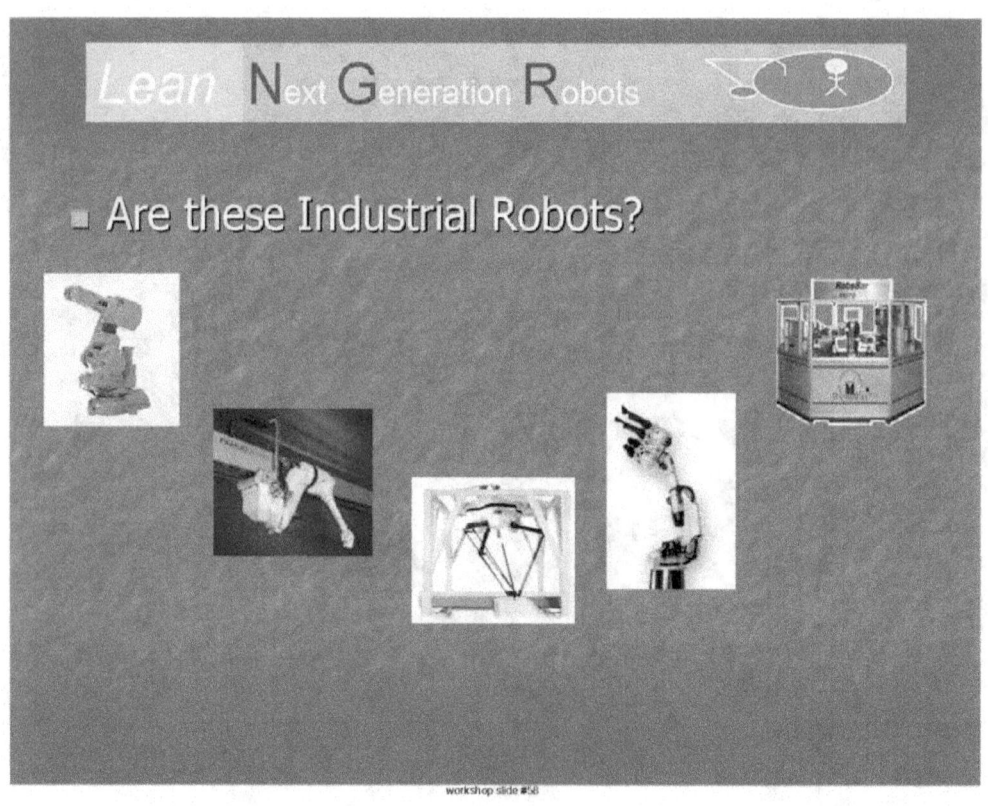

Lean Next Generation Robots

What is a lean Next Generation Robot?

- The main characteristic of lean Next Generation Robots is that they will be capable of working more closely with Humans

- ANSI/RIA/ISO 10218 Standards are in development to safely enable this functionality which is called "Collaborative Operation"

Lean Next Generation Robots

Which installation is safe?

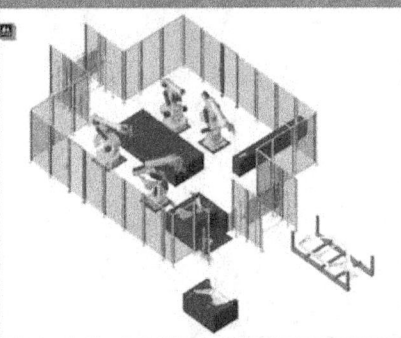 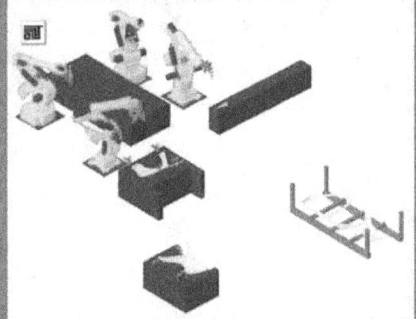

With future NGR technology, the answer may not necessarily be the one with the highest fence!

Lean Next Generation Robots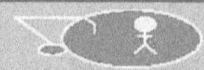

Evolution of Industrial Robot Safety

- Early 1980's – No safety standards specific to robots
- 1986 – First RIA 15.06 Safety Standard Published, revisions – 1992 & 1999
- 2007 - ANSI/RIA/ISO 10218-1 Adopted (August)
- However, as this level of safety has been raised, the amount of restrictions have increased. Basically the direction has been to prevent close human interaction with the robot whenever possible.
- In these early years "safe" meant higher fences, more E-Stops and more ways to keep humans away from the automation. Increased clearances distances, more floor space required, more fencing, more complex and more costly.
- This trend is restrictive, very expensive and inefficient!

Lean Next Generation Robots

- Safe meant <u>stop</u> (requiring a fault clearance and manual restart) whenever a human was too close; But, is this best for productivity?
- Is it best to stop the robot with a binary decision just because someone is near it?
- What if the robot slowed down when someone was close and stopped when they are too close and in danger?

Lean Next Generation Robots

NGR Technology

- In the next several years robot controllers will not only have control reliable, redundant stopping circuits; but they will also have control reliable feedback for axis position and motion is used as well.

- Software logic can be used for safety functions that were formerly limited only to hardware circuits.

- Software logic can be used for safety functions as long as it is protected from reprogramming, redundant and cross checked with multiple CPU monitoring. (This technology is used today in Safety rated PLCs)

Lean Next Generation Robots

"Collaborative Operation"

- Newly released (August 2007) ANSI/RIA/ISO 10218-1 standard defines Collaborative Operation Requirements in Clause 5.10

    - 5.10.2 - The robot shall stop when a human is in the collaborative workspace...

    - 5.10.3 - When provide, hand guiding equipment shall be located close to the end-effector and shall be equipped with: a) an emergency stop and b) an enabling device...

    - 5.10.4 - The robot shall maintain a separation distance from the operator (necessitating detection by a visual/optical means); This distance shall be in accordance with ISO 13855. Failure to maintain the separation shall result in a protective stop...

Lean

Lean means to eliminate waste. For robot cells and lines lean means:

- Decreased Floor Space
- Safer, yet lower cost safety integration
- Faster Integration/Startup Time
- More Flexibility
- Lower operating cost
- Increased uptime - High MTBF, Low MTTR

Benefits of Lean NGR

- Decreased floor space
- Increased productivity
- Reduced waste
- Reduced guarding
- Zero clearance
- Improved performance
- Safety improvements that allow new applications ie. collaboration with human activities

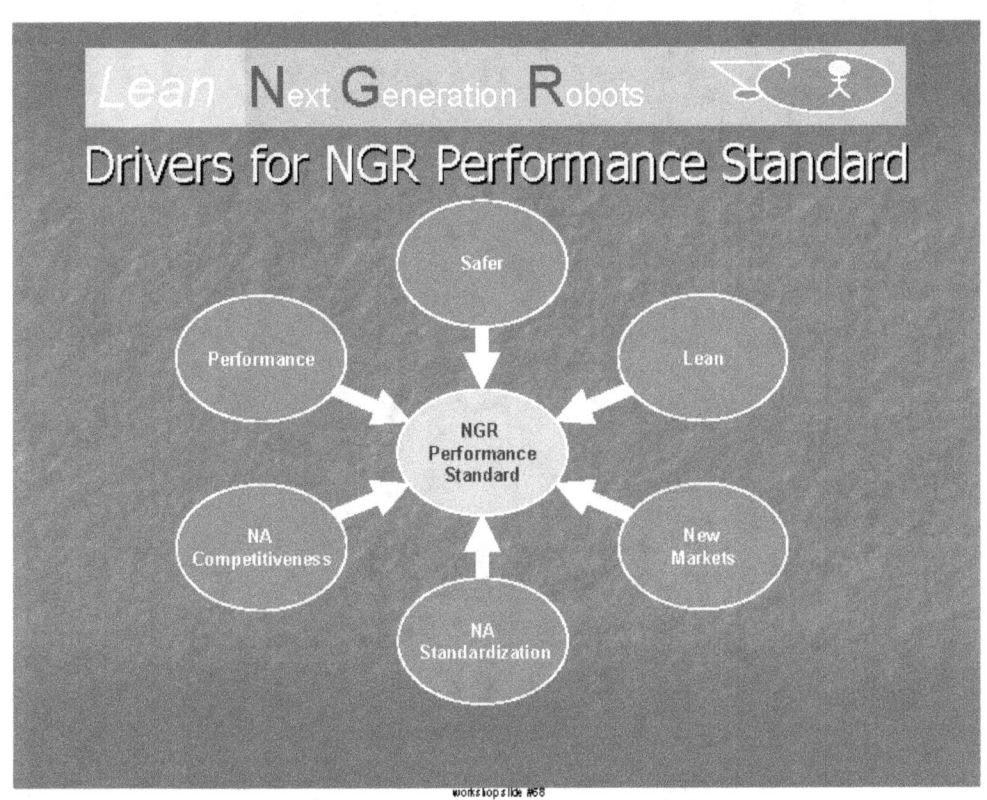
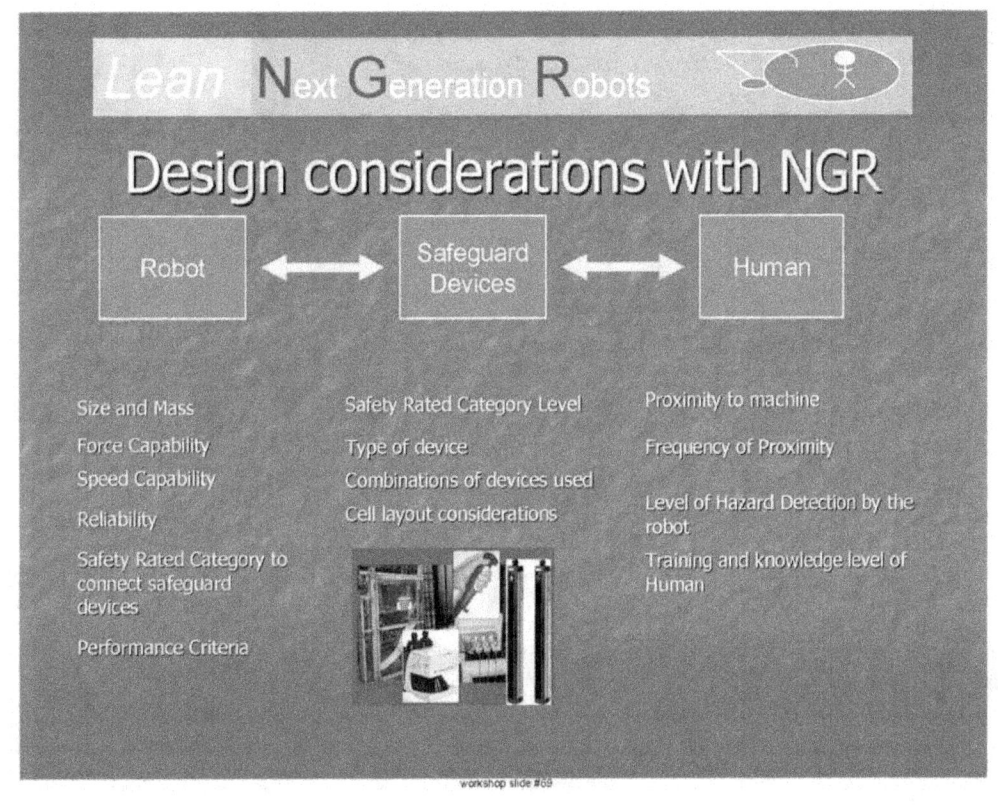

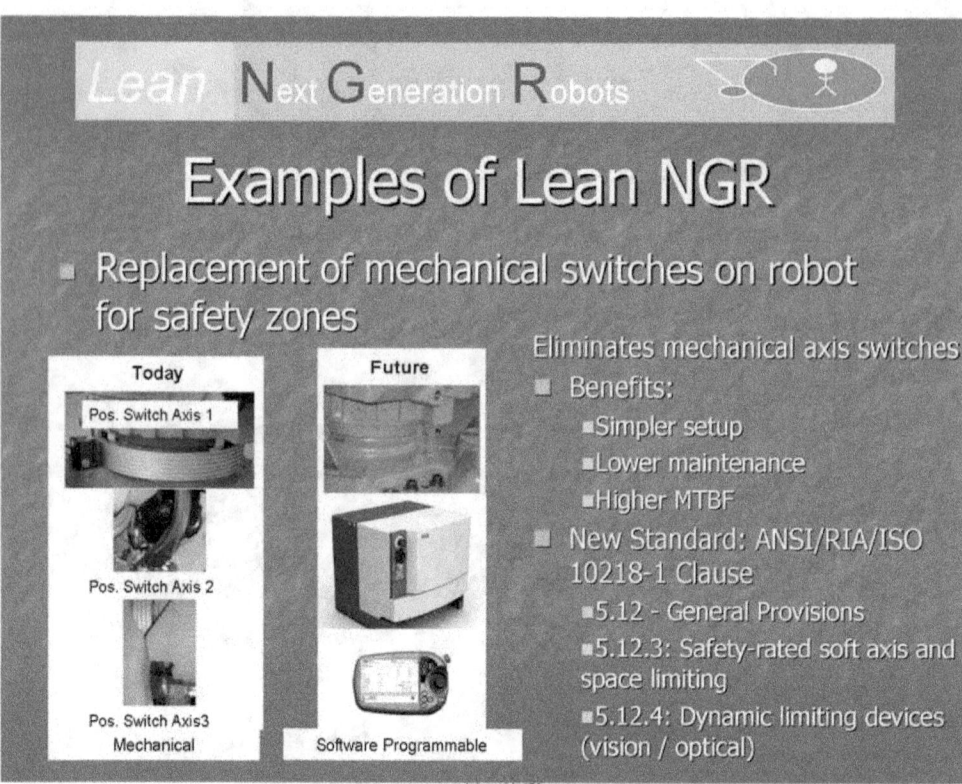

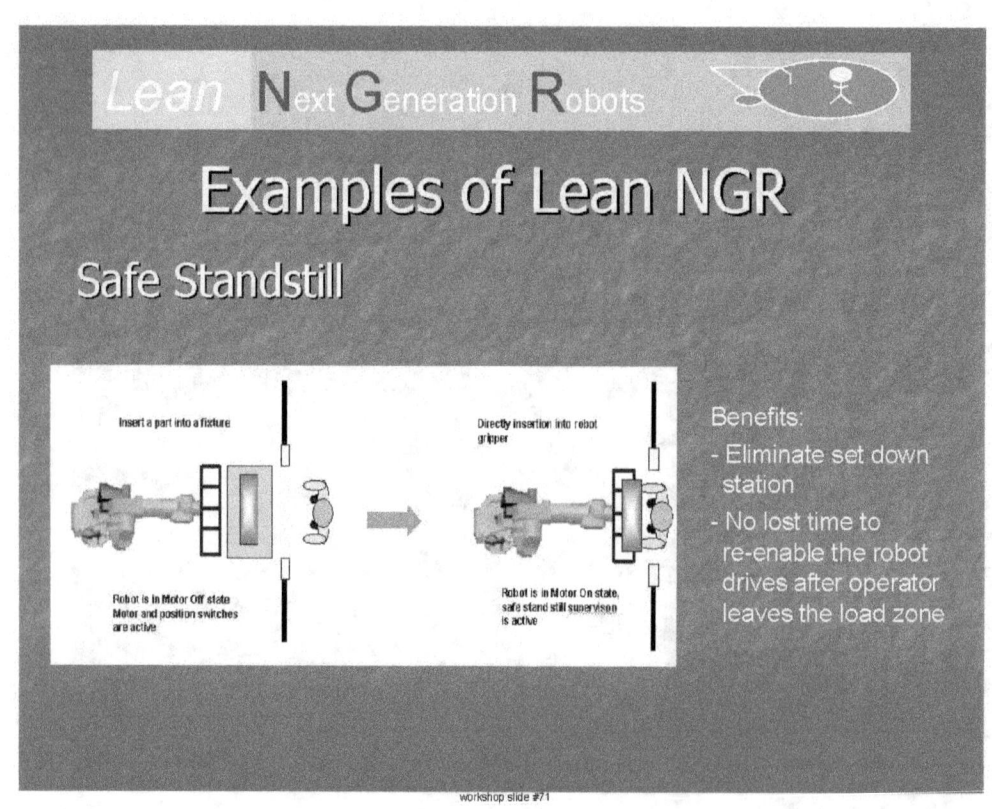

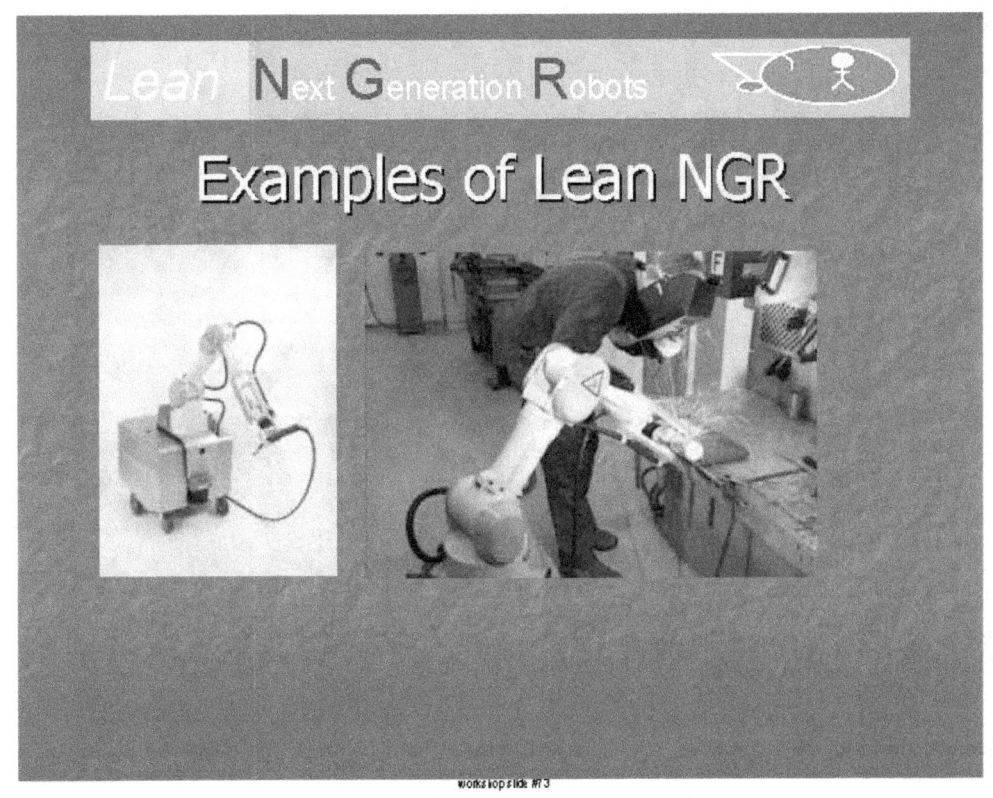

Lean Next Generation Robots

Concept to categorize robots based on Safety Features

	Old Safeguards			Human Proximity Safeguards			
Industrial Robot Safety Category	1	2	3	4	5	6	7
Emergency Stop	X						
Stopping Circuit		X					
Brakes		X					
Adustable Hard Stops		X					
Brake Release		X					
Hold to Run Switch		X					
Slow Speed Teach		X					
Enabling Device			X				
Control Relabile Stopping Circuits			X				
Control Relabile Axis Range switches (DLD)			X				
Redundant Feedback (RF) Range limit				X			
Redundant Motion detection capabiltiy				X			
Proximity Slow Speed Auto Mode				X			
Safe I/O / Field Bus					X		
Auto Mode Safe Speed Limit (RF)						X	
Auto Mode Safe Accel/Decel Limit (RF)						X	
Lowered Torque limits						X	
Human proximity Detection							X
Human Position avoidance							X

(Today / Future)

Lean Next Generation Robots

Conclusions

- NGR technology is in our future
- NGR technology has promise to be safer and leaner than present technology
- Performance and Safety Standards will need to be written to accommodate the new technology
- Robotics industry manufacturers, integrators and users will need to understand the capabilities

Lean Next Generation Robots

RIA actions

- Work Group formed to begin discussing a framework of Lean NGR Performance standard.

- Awareness presentations to stimulate interest and acceptance within working groups representing manufacturers & users.   RIA, AIAG, USCAR etc.

Mobile Manipulation:
Going beyond AGVs

George Kantor
kantor@cmu.edu
The Robotics Institute,
Carnegie Mellon Univeristy

in collaboration with:
Sanjiv Singh
Seth Koterba
Brad Hamner
D.H. Shin
M. Hwangbo

Dynamic Measurement and
Control for Autonomous
Manufacturing Workshop
10 October 2007

AGV Overview

- Automated Guided Vehicles (AGVs) follow prespecified guidepaths
- Wide range of material handling applications
- Many vehicles forms (forked, tow, loader)
- Many manufacturers (AGV, FMC, Savant, Webb, Egemin, COH)

Key Issue: Localization

- **Wired** -- guidepath defined by wires embedded in floor
- **Inertial** -- localization from magnetic beacons embedded in floor
- **Laser** -- localization using laser scanner and reflectors
- **Visual** -- localization using on-board cameras (e.g., SEEGrid)

Manipulator Overview

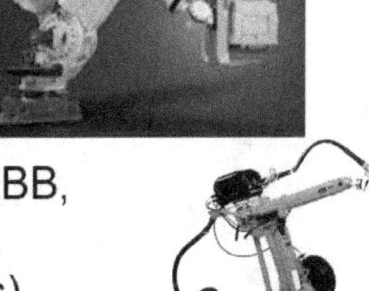

- Robot arms with fixed workcells
- Many applications
- Over 30 years of commercial success
- Many manufacturers (ABB, FANUC, KUKA, Denso, Unimation, many others)

Key Issue: Manipulability

- Manipulability affects:
  - Accuracy
  - Strength
  - Speed
- Is a function of configuration
- Must be considered when planning tasks

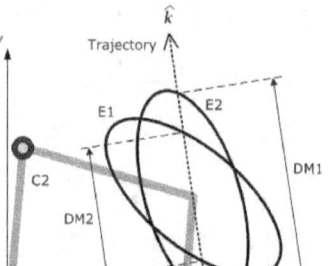

C: Joint Configuration(elbow-up, down)
E: Velocity Manipulability Ellipsoid
DM: Directional Manipulability

Mobile Manipulation

- Large workcell size
- Reduced infrastructure
- Flexible manufacturing

Key Issue: Accuracy

accuracy = localization + manipulability

Related Work

- Coordination of locomotion and manipulation
  - Redundancy optimization: Carriker
  - Maximizing manipulability: Yamamoto
  - Compensation of the dynamic interaction of the base and the manipulator:
    - Tip over: Huang and Sugano
    - Vehicle suspension: Hootsmans
- Cooperation of multiple mobile manipulators
  - Derived from the force control methodology
- Control execution (RMRC for mobile manipulator)

A First Approach

Selecting base poses:

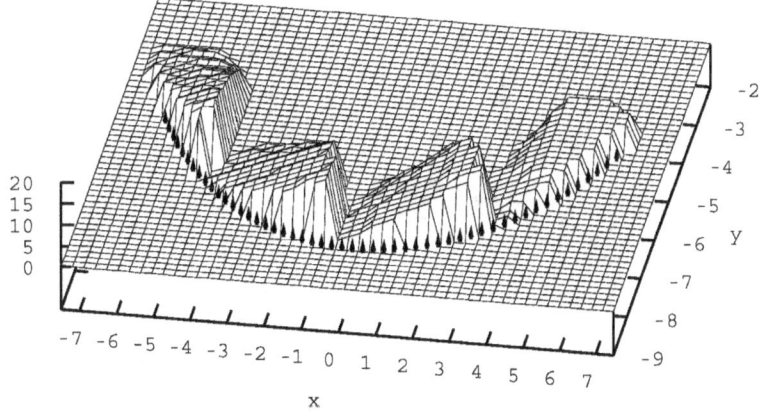

Each grid cell gets a score based on **how much** of the path and **how well** the it can be covered with the base at that point.

Example: 2 link manipulator

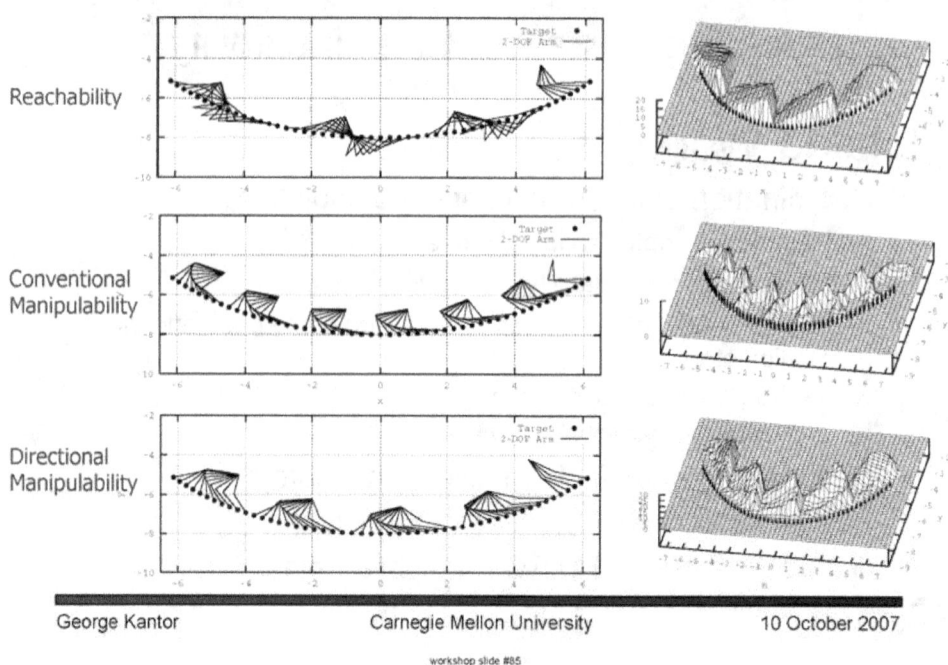

Reachability

Conventional Manipulability

Directional Manipulability

Experimental Results

Current/Future Direction

Integrated base/manipulator motion

QuickTime™ and a
Cinepak decompressor
are needed to see this picture.

A Different Direction for AGVs

QuickTime™ and a
Cinepak decompressor
are needed to see this picture.

A Different Direction for AGVs

QuickTime™ and a
Motion JPEG OpenDML decompressor
are needed to see this picture.

George Kantor — Carnegie Mellon University — 10 October 2007

Dynamic Visual Servoing

Avi Kak
Johnny Park
German Holguin

Robot Vision Lab
Purdue University

PURDUE
UNIVERSITY.

Purdue Robot Vision Lab

- Founded in 1978

- Performs state-of-the-art research in sensory intelligence for the machines of the future

- Produced 39 Ph.D.'s so far

Significant Industrial Collaborations

- Olympus Corporation (2004 ~ present)
- Infosys Technologies (2007 ~ present)
- Ford Motor Company (1994 ~ 2006)
- Mitsubishi Heavy Industries (1992 ~ 1993)
- Denso Corporation (1991 ~ 1995)
- Hitachi Heavy Industries (1991 ~ 1997)

Relevant Research Projects

- 3D Object Recognition and Bin Picking
- Vision-Guided Mobile Robot Navigation
- 3D Modeling
- Real-Time Background Subtraction in Video Imagery
- Distributed Sensor Networks

- **Line Tracking for Assembly On-the-fly**

Line Tracking for Assembly-on-the-fly

- Goal:
  - Develop a vision-guided robotic system that can operate on a moving assembly line
  - Replace "stop-stations" in the assembly line

Line Tracking System (video)

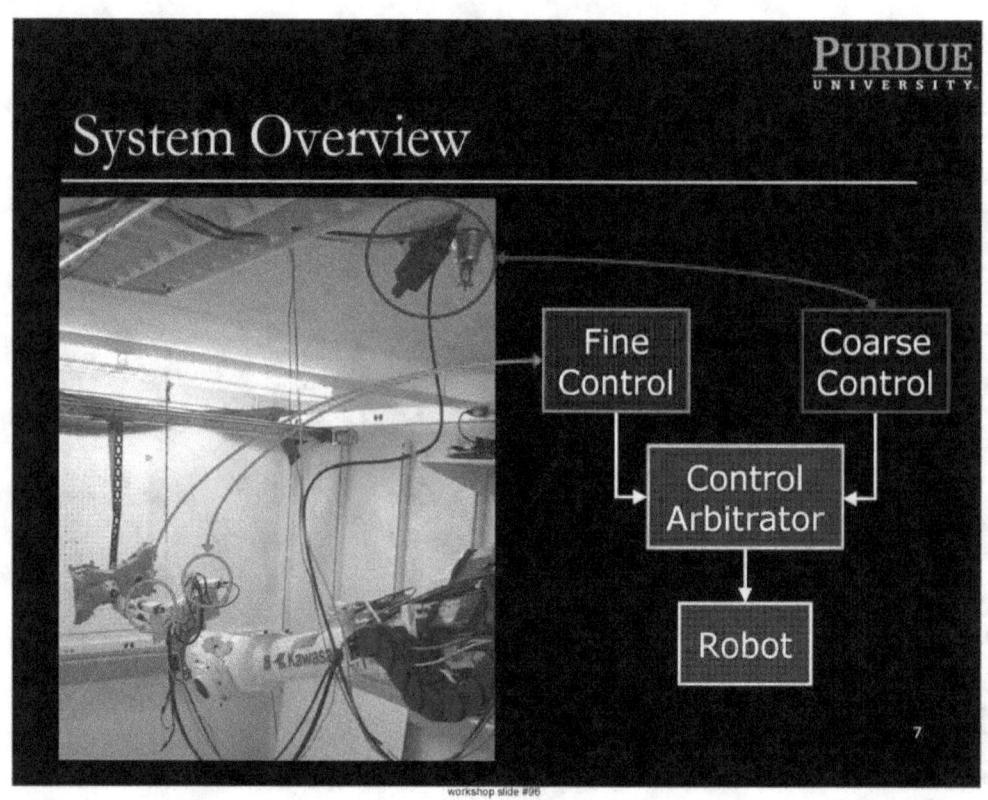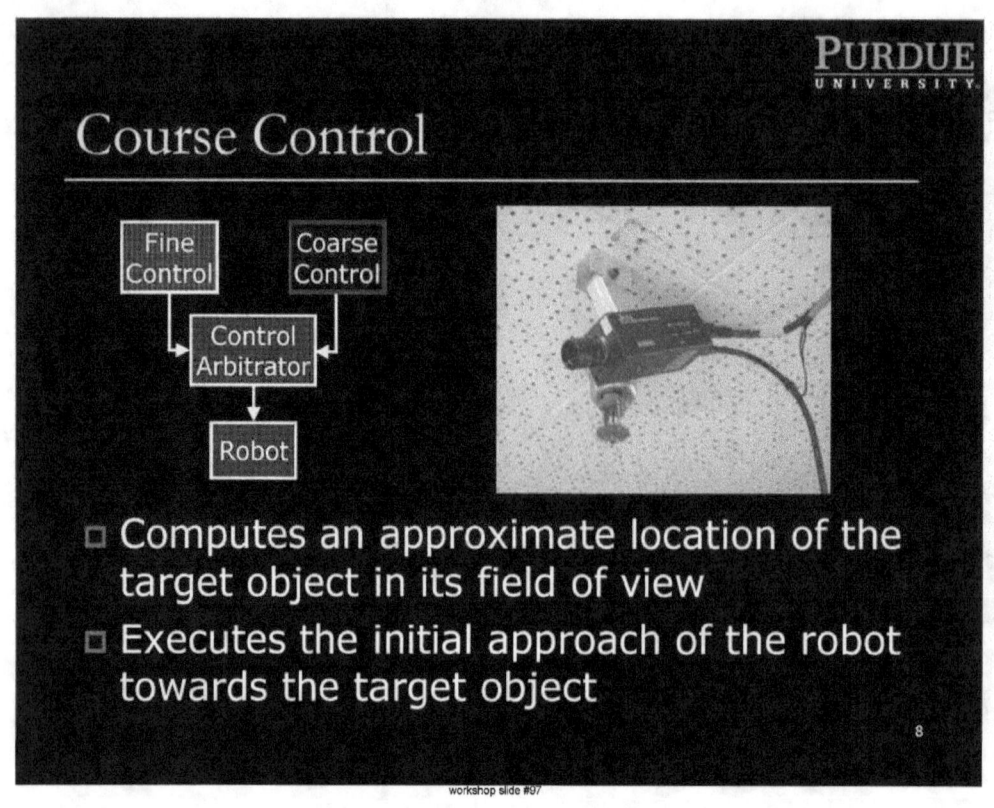

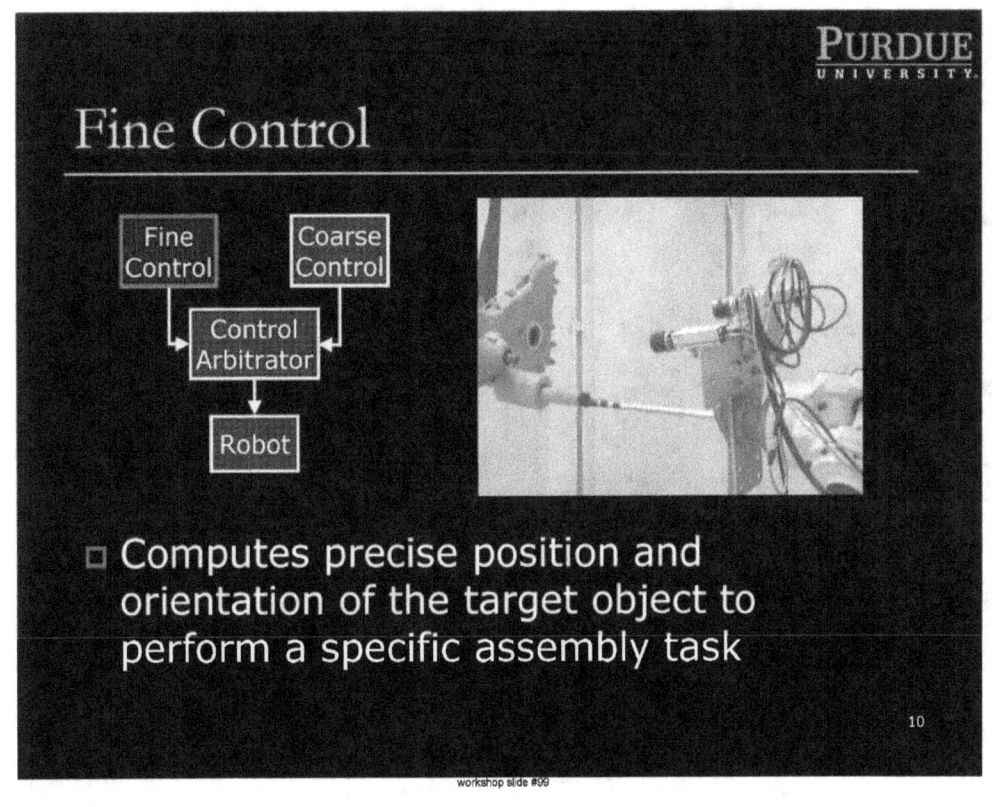

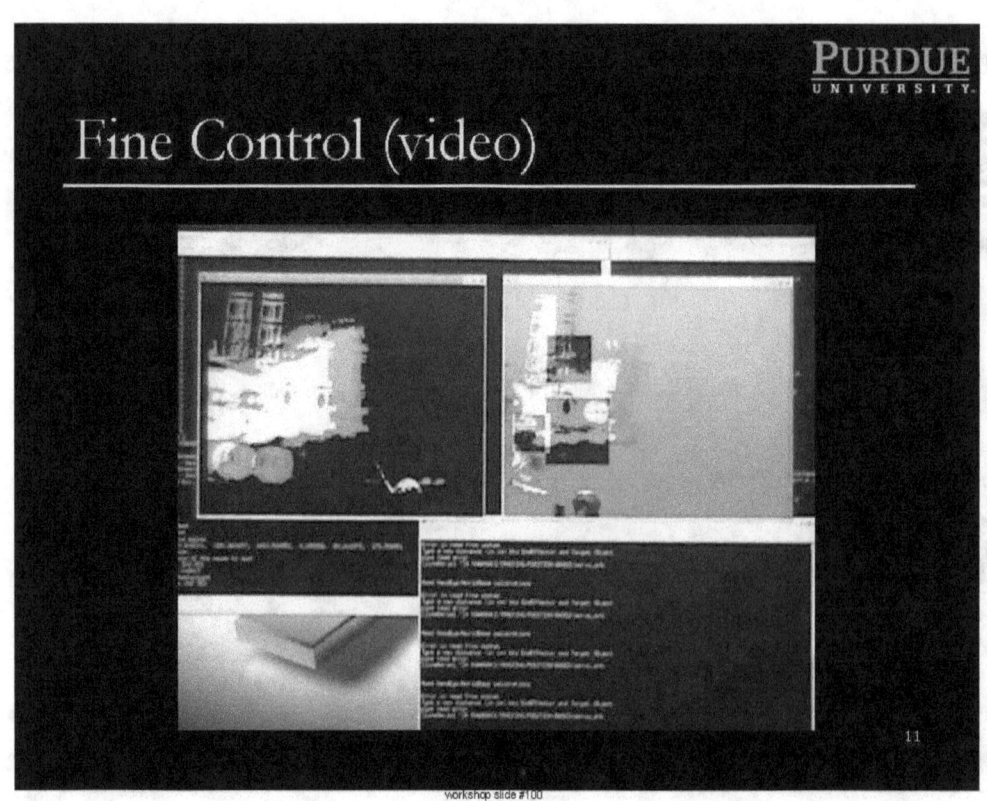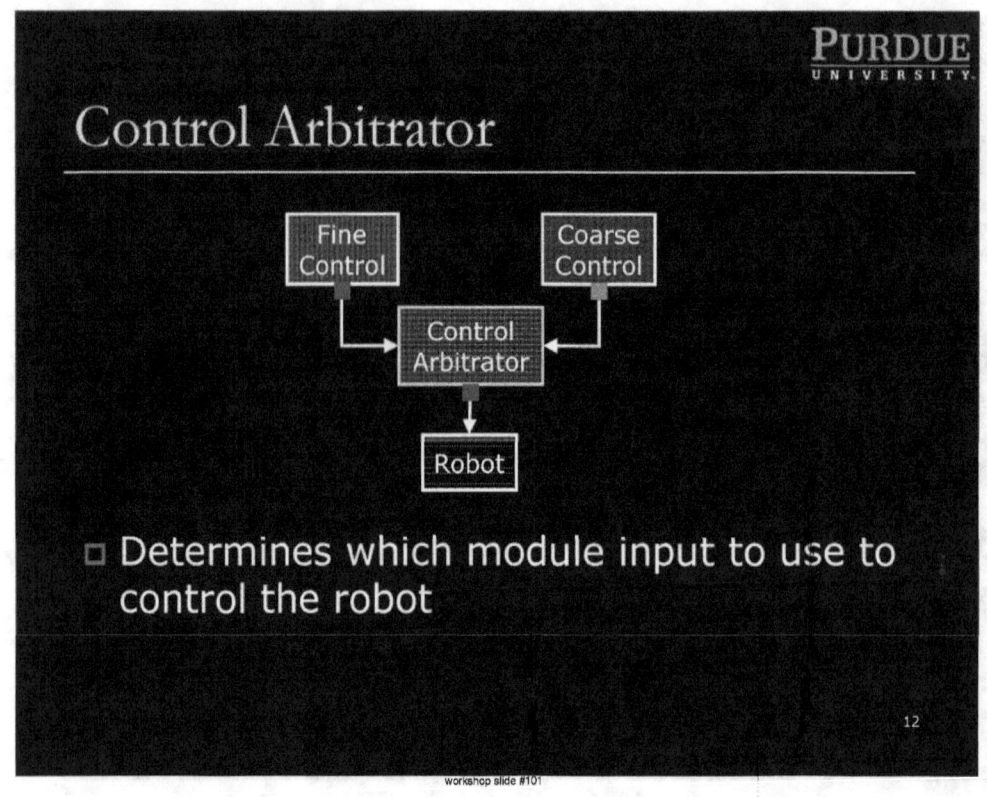

Line Tracking System (video)

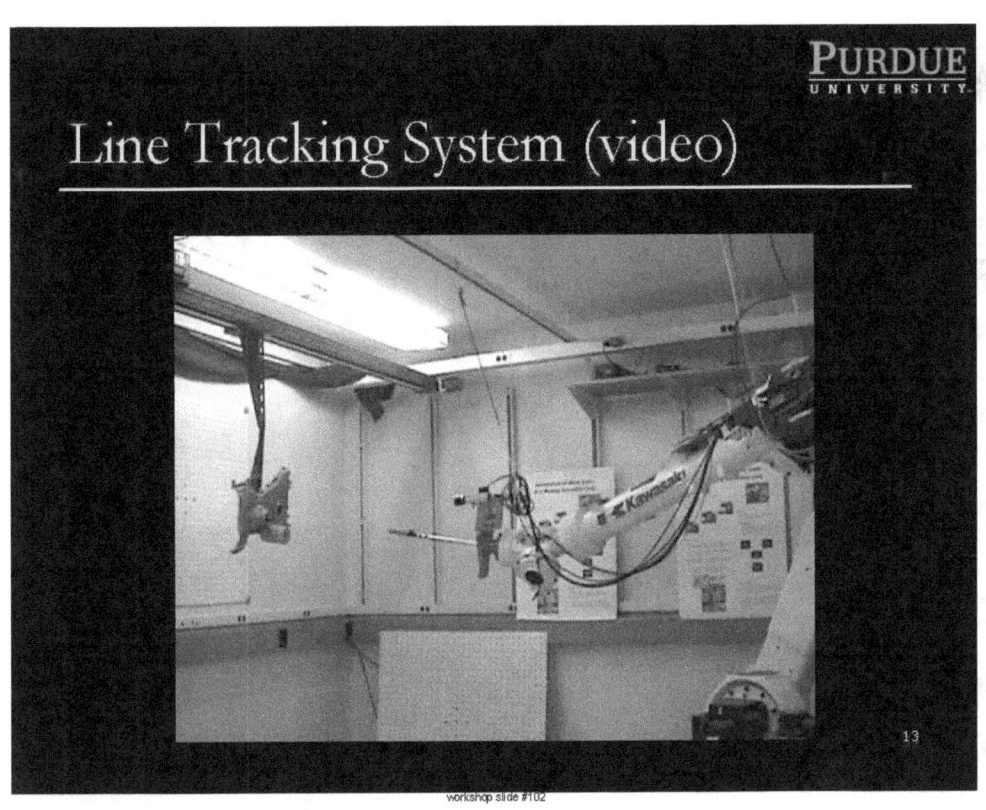

Distributed and Hierarchical Control Architecture

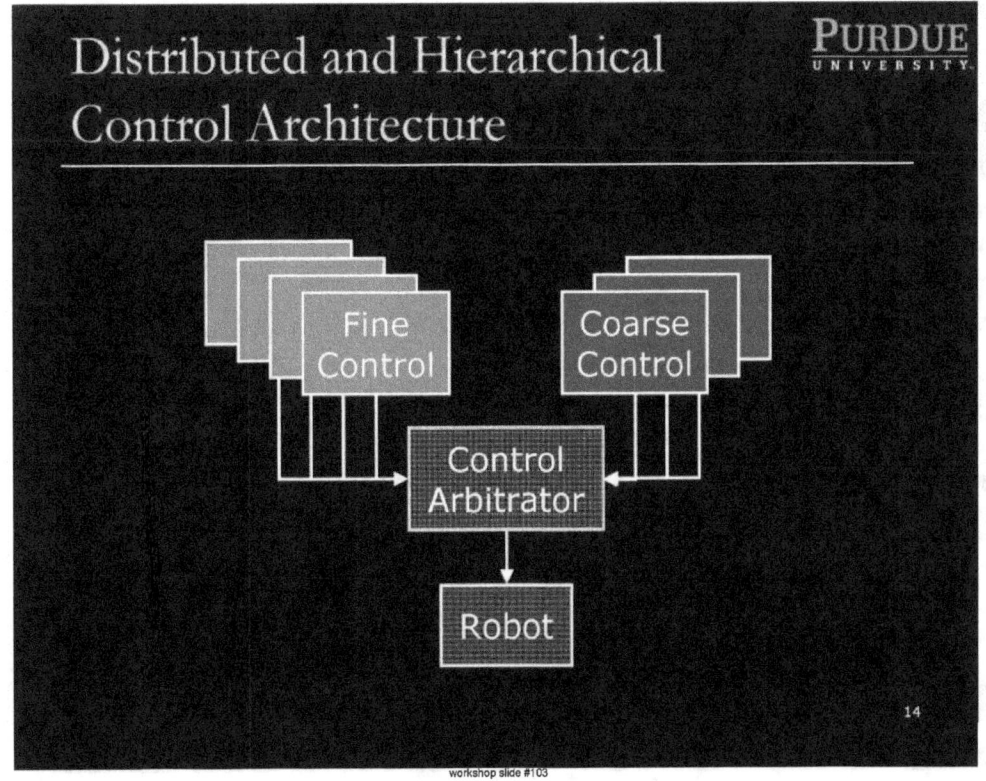

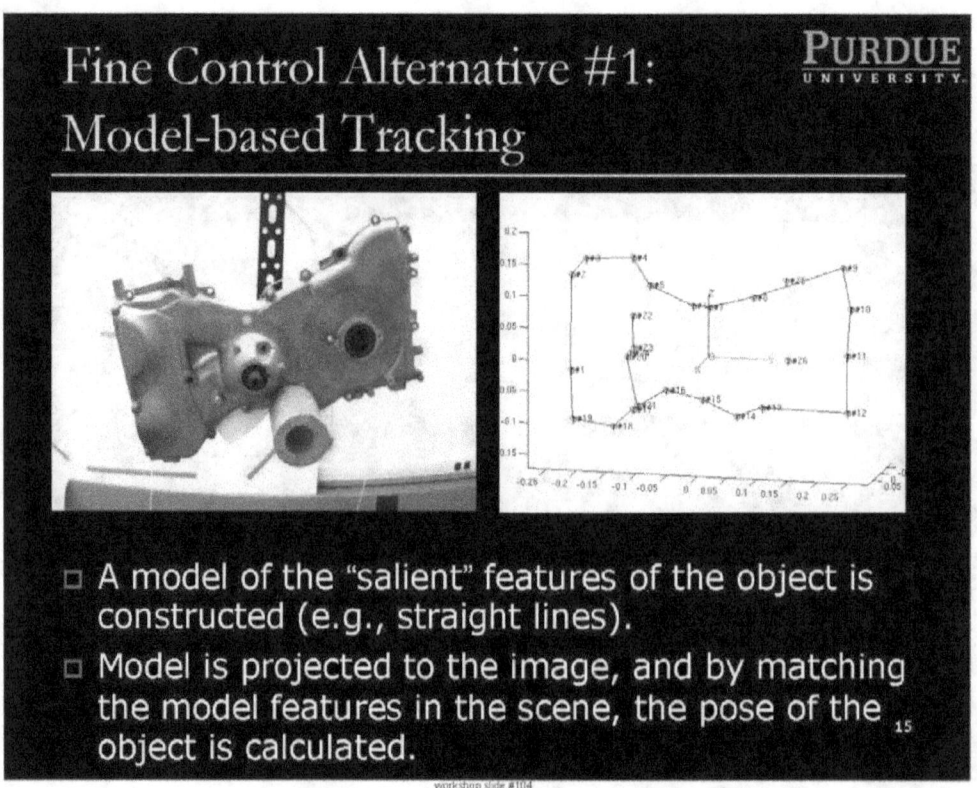

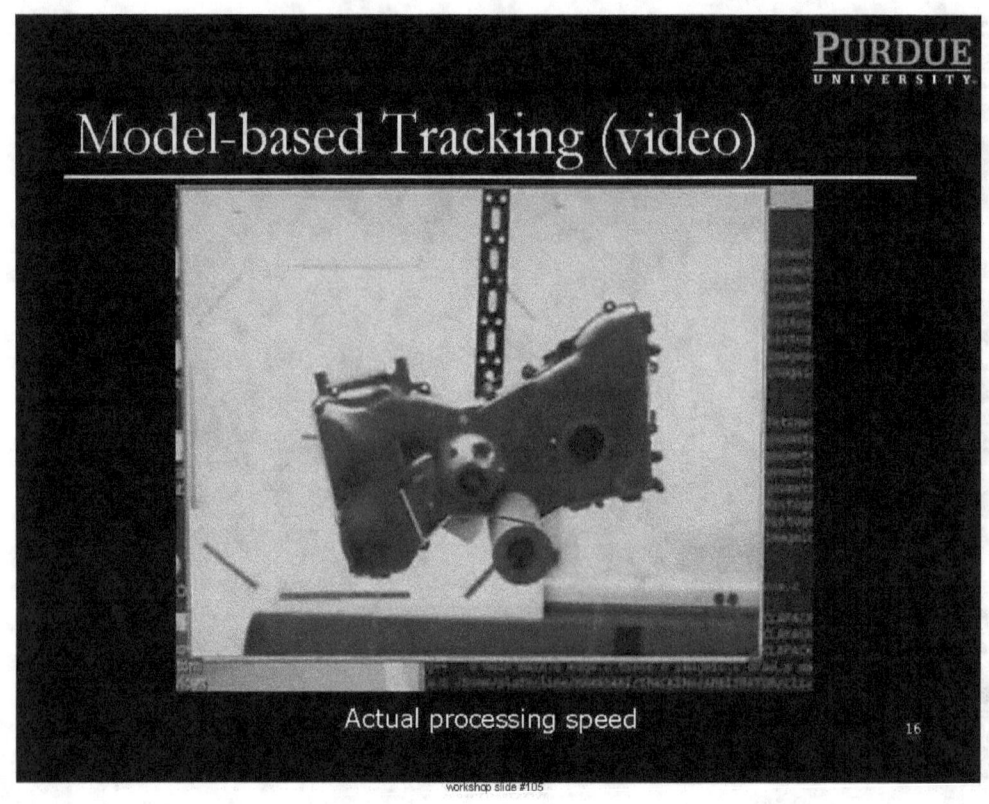

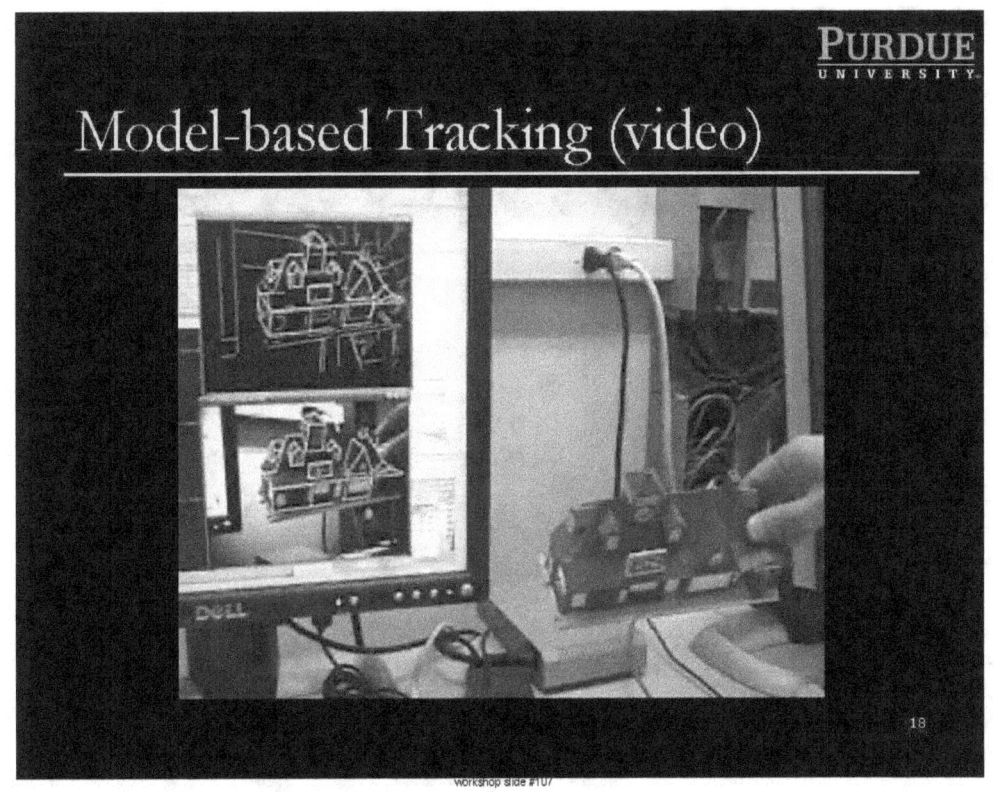

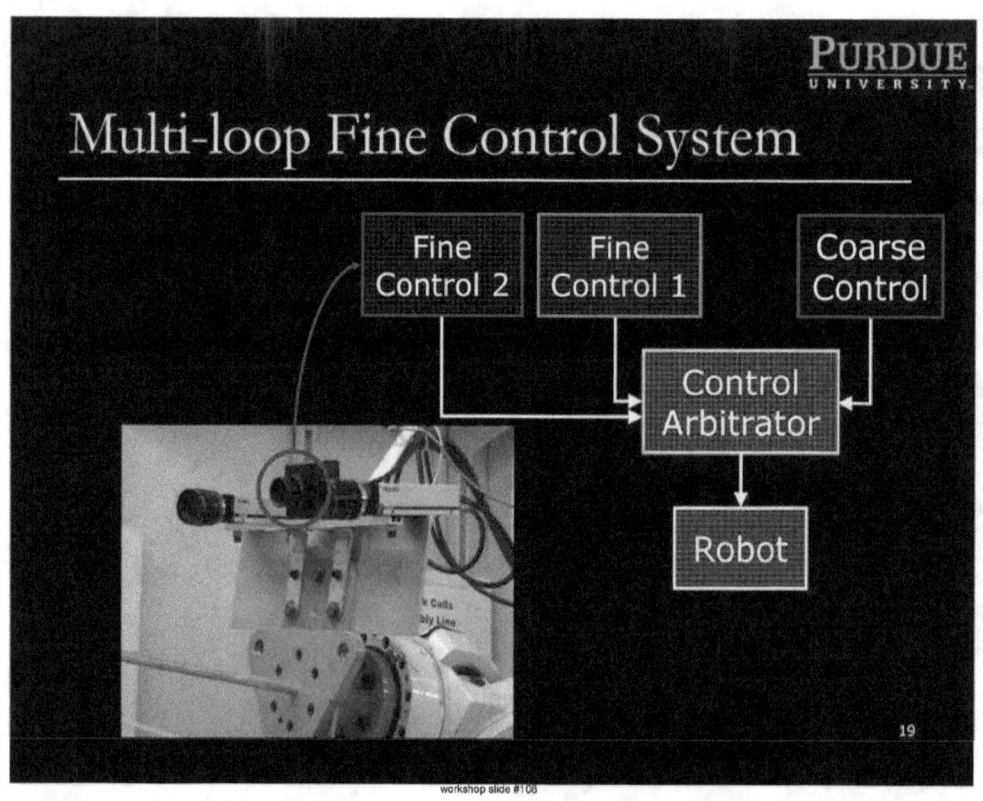

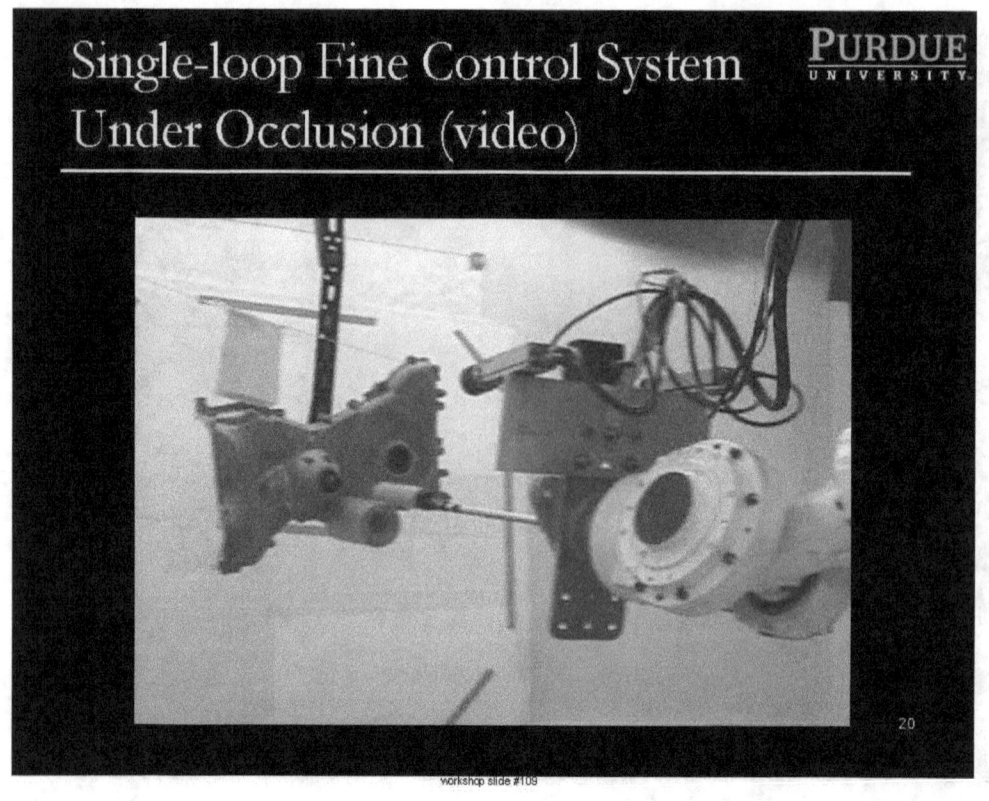

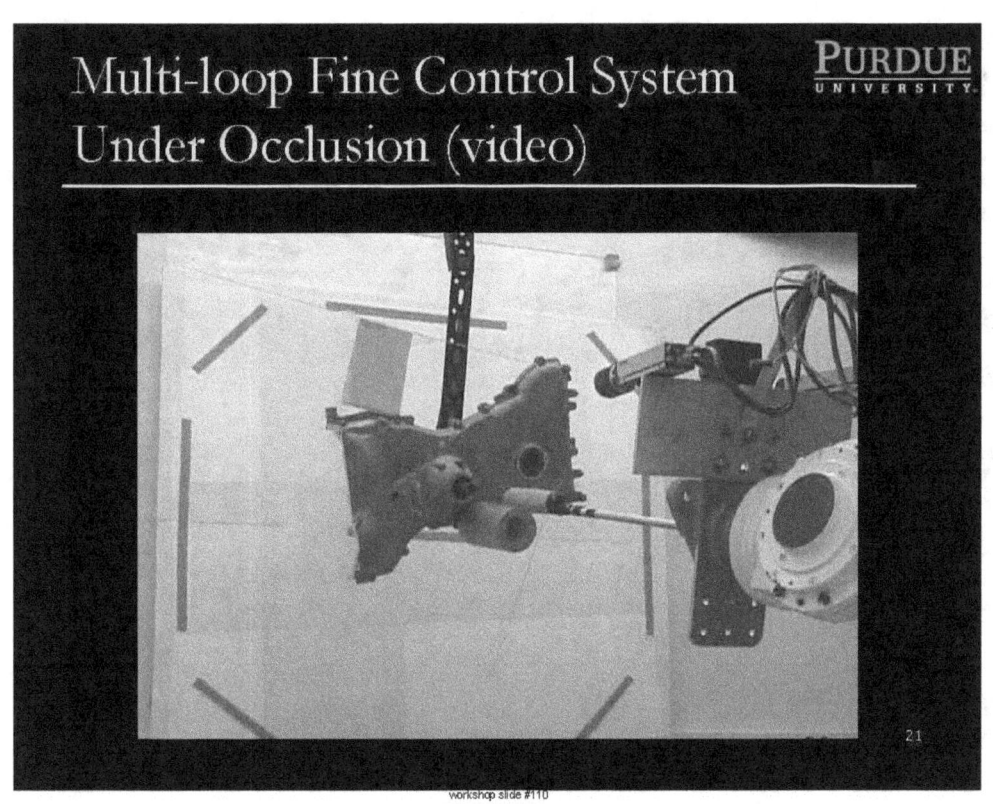
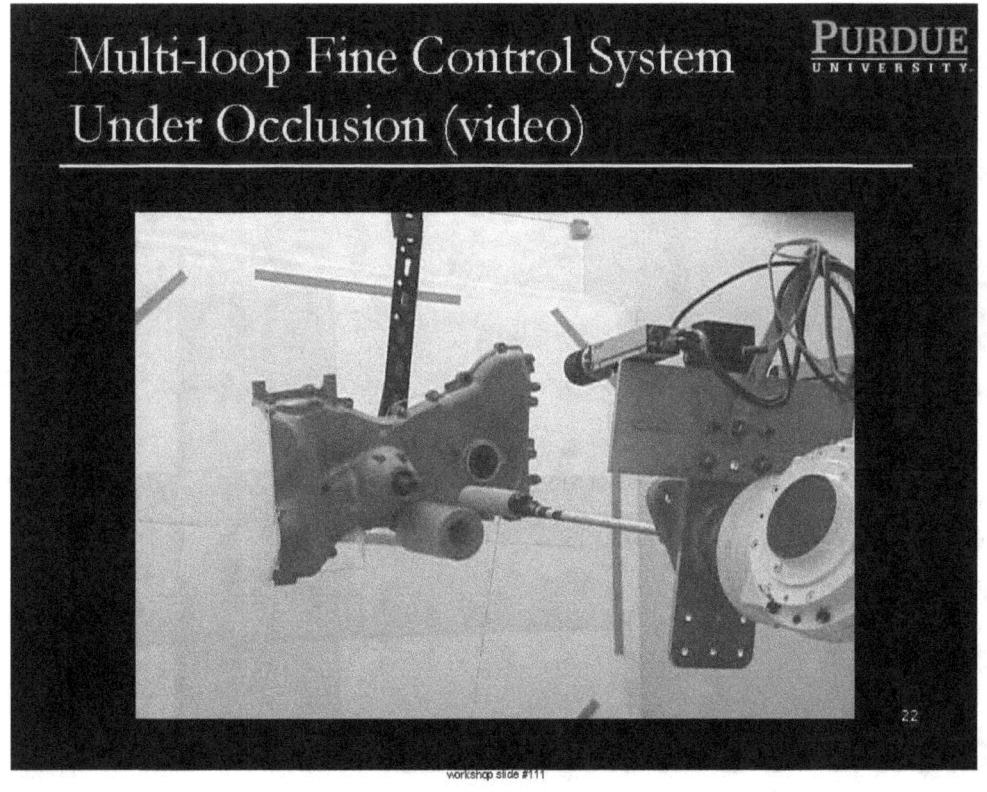

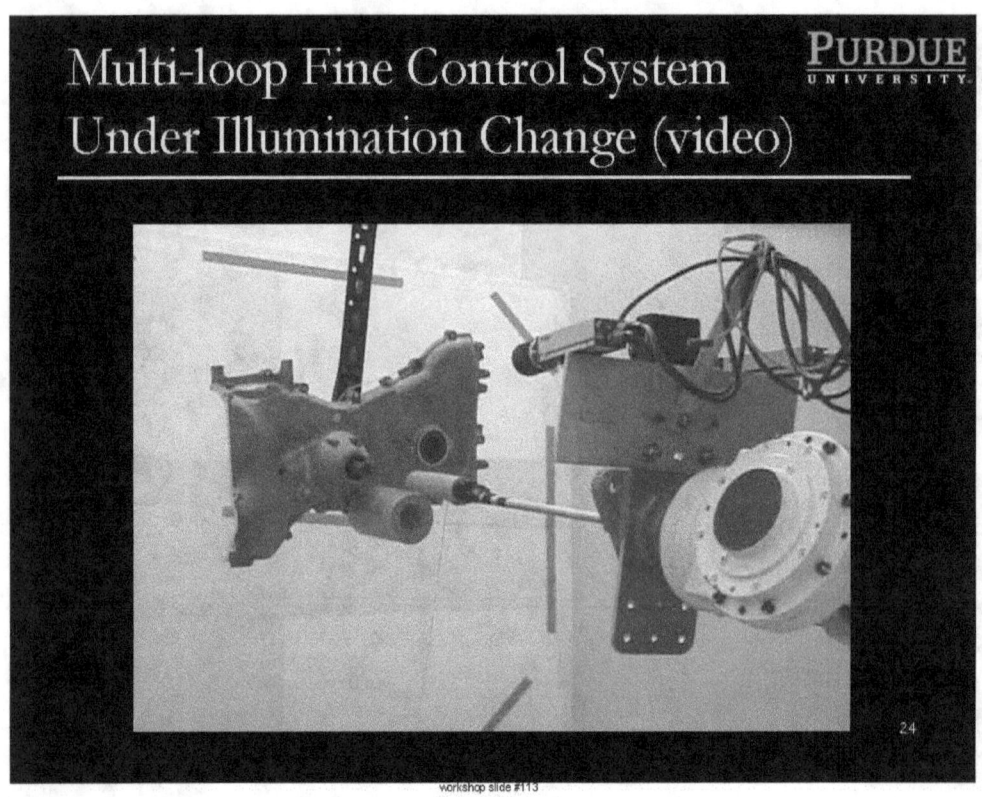

For More Information

- Please visit our website:

  http://rvl.ecn.purdue.edu

Visual SLAM:
Past, Present & Future

Motilal Agrawal, Kurt Konolige, Joan Sola

SRI International

NIST Workshop on
Dynamic Measurement and Control for
Automated Manufacturing

October 10-11, 2007

www.ai.sri.com/~agrawal,konolige

Overview of the talk

- Introduction to Visual Odometry & SLAM
- History
- Visual Odometry Principle
- Current Status and Directions
- Results on various datasets

Visual Odometry & SLAM

10/09/07

- ➢ Task
  - **VO:** estimate the pose of a vehicle
  - **SLAM:** Build maps and stay localized in this map
- ➢ Sensors
  - **Accelerometers/IMU** accumulate error rapidly
  - **Wheel Odometry** is subject to slip, sliding
  - **GPS** (WAAS) is accurate to 3-5 m in the best case in open outdoor terrain; is worse under tree canopy, inside buildings and is subject to jamming;
  - **Visual Odometry (VO)** has tremendous potential
    - Can complement other sensors
- ➢ Applications
  - Estimating 6 DOF pose of objects on the assembly line
  - Estimating the pose of a robot indoors
  - Autonomous navigation

History

10/09/07

- ➢ Cameras are cheap now
  - Stereo cameras work best for visual SLAM
- ➢ Computing power has gone up
  - Specialized hardware for stereo exist
- ➢ Vision algorithms for structure from motion are now viable
  - Maturity in Computer Vision
- ➢ A few systems around the globe for real time visual SLAM
  - SRI has been developing Visual Odometry for three years now

VO principle

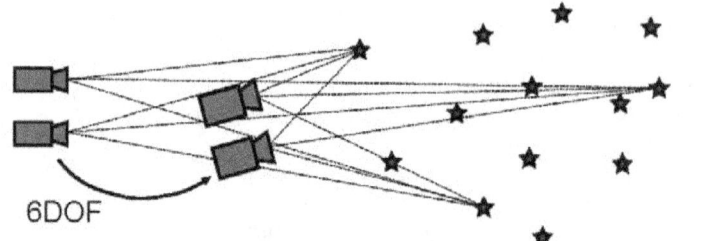

Images with 3D information from stereo

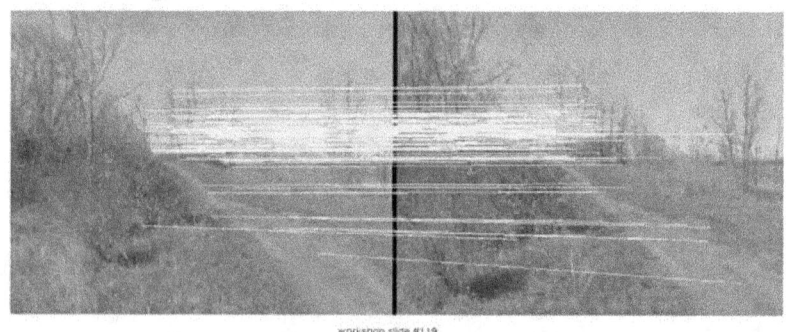

VO provides 6DOF relative motions

Use of Key frames and Window mesh reduces drift

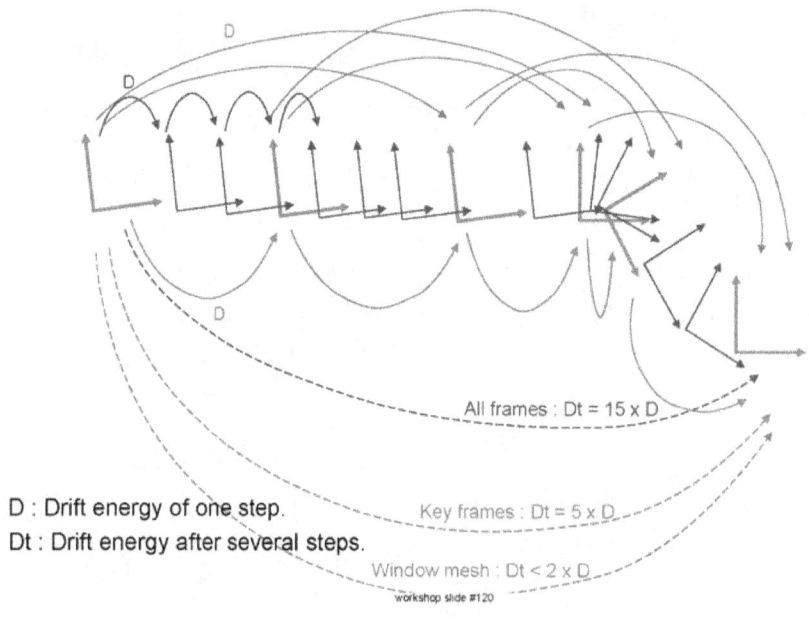

D : Drift energy of one step.
Dt : Drift energy after several steps.

All frames : Dt = 15 x D
Key frames : Dt = 5 x D
Window mesh : Dt < 2 x D

Indoor Feature Tracking

Indoor office environment

VO in an indoor office corridor

Loop of 40 meters

Loop closure error < 0.4 m

Crosses show the trajectory
    forward trajectory in blue
    return trajectory in red
The 3D tracked features are marked as dots

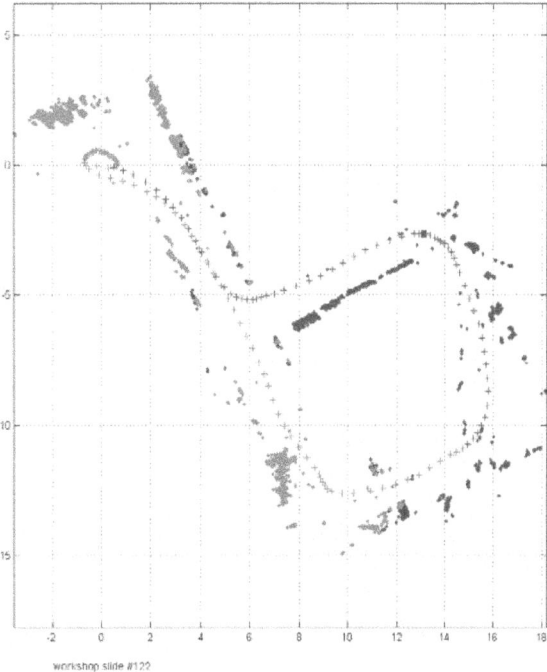

Current status

- Practical, inexpensive, real-time vision based system for localization
- Localizes within 1% error over large distances
  - Experimental validation over 9km on outdoor terrain
- System ideal for autonomous navigation of a robot
- Long term drifts minimized through integration with a low cost absolute sensor
  - Gravity normal from IMU
- System tested out on datasets from other people

Ongoing work

- Good features for tracking
  - Indoor vs outdoor
- SLAM and loop closure
  - Maps remain consistent in spite of drift
- Visual landmarks recognition
  - Relocalization
- Integration with IMU at the sensor level
  - IMU provides an absolute gravity normal to provide the angle corrections
- Visual SLAM workshop
  - IROS 2007, Nov 2, 2007
  - CVPR 2008, June 23 2008

Results on datasets provided by other people   10/09/07

 Visual Odometry Example: Urban environment   10/09/07

- Thanks to Andrew Comport, LAAS, CNRS France
- Outdoor sequence in Versaille
- 1 m stereo baseline, narrow FOV
- ~400 m sequence
- Average frame distance: 0.6 m
- Max frame distance: 1.1 m

10/09/07

Visual Odometry Example

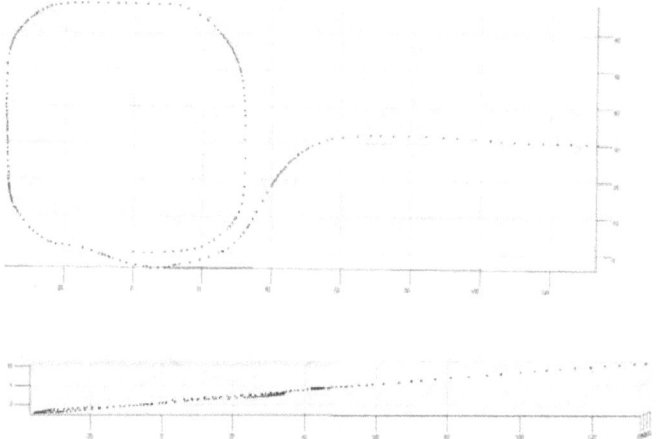

- Outdoor sequence in Versaille
- 1 m stereo baseline, narrow FOV
- ~400m sequence
- XY and Z plots of each frame
- STD Z error to a plane fit is 14 cm

 10/09/07

Visual Odometry Example: Indoor lab sequence

- Thanks to Robert Sim, UBC, Canada
- Indoor lab sequence
- 12 cm stereo baseline, wide FOV
- ~100 m sequence, ~1200 key frames
- 17 tack points in the VSLAM graph

Visual Odometry Example

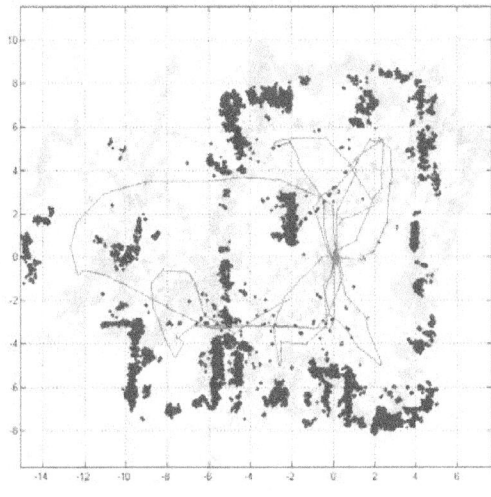

- Indoor lab sequence
- 12 cm stereo baseline, wide FOV
- ~100 m sequence, ~1200 key frames
- Green crosses are uncorrected VO; cyan environment points
- Red segments are VSLAM-corrected poses; blue environment points
- 17 tack points in the VSLAM graph

VO on a very large outdoor sequence
Raw VO compared to RTK GPS (ground truth)
9 km path length
50 m end error (0.5%)

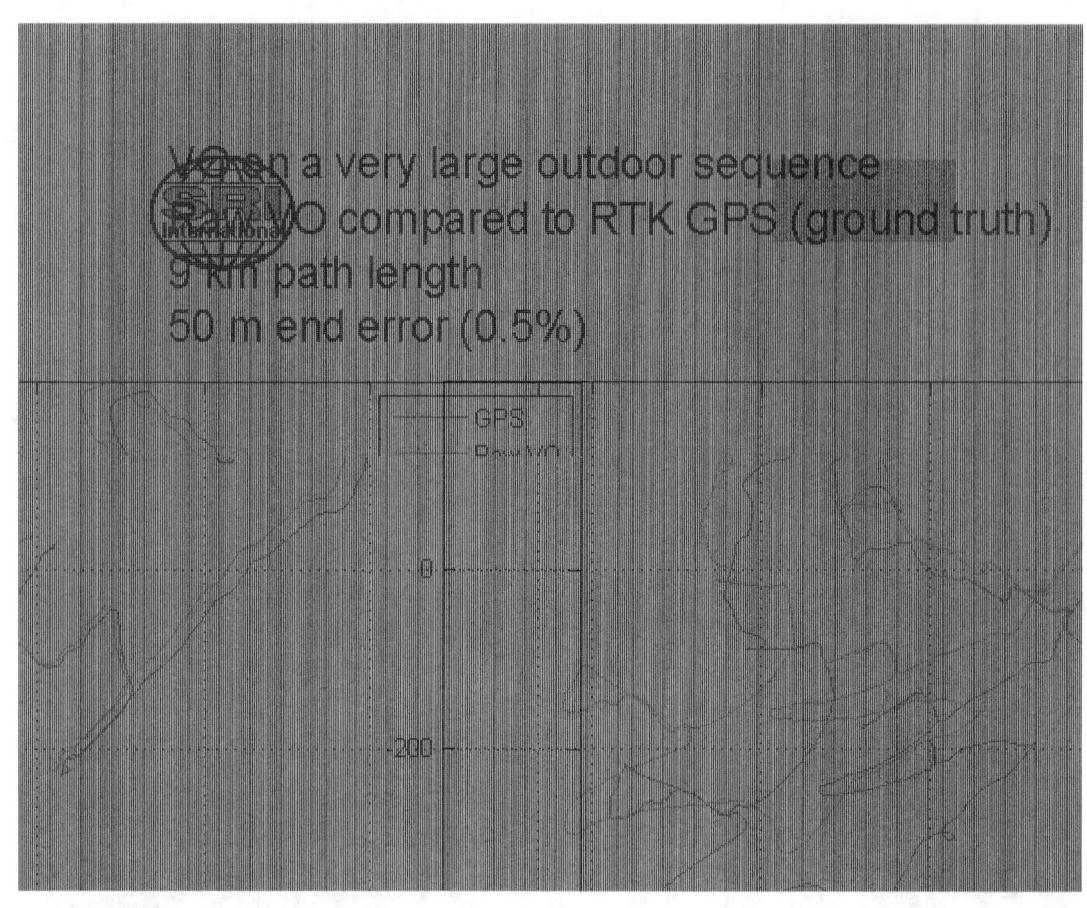

Fusion with IMU using Kalman Filter
10 m end error (0.1%)

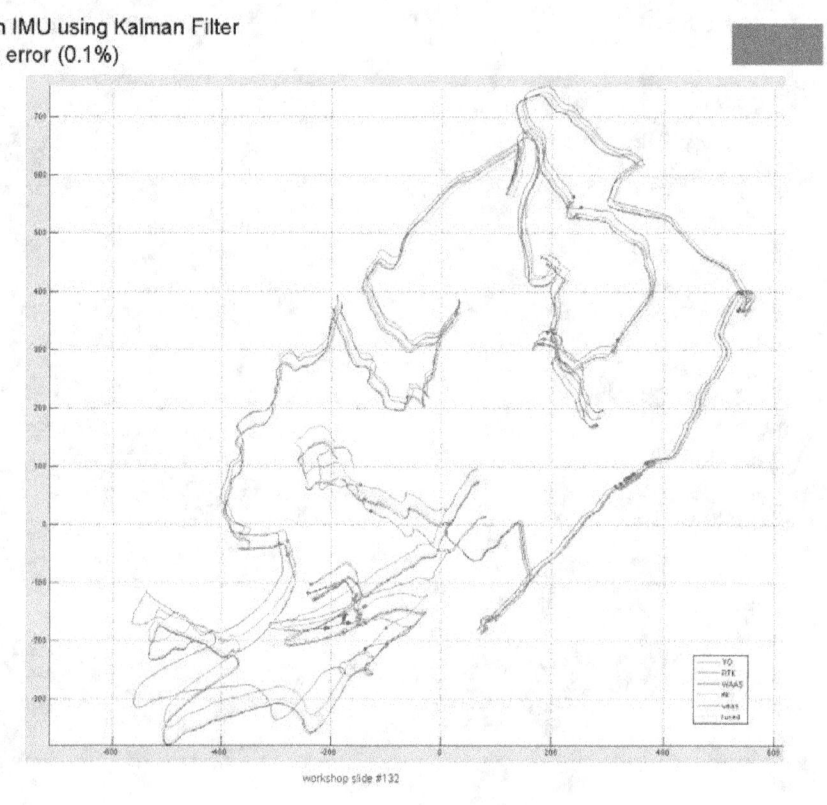

ASTM E57 on 3D Imaging Systems
Alan Lytle

- A proven and practical system
  - Established in 1898
  - 140 Committees & 12,000+ Standards
  - 30,000 members
    - 5,500+ International Members from 125 countries
    - 3,000 ASTM standards used in 60+ countries
  - 'Audited Designator' accreditation: American National Standards Institute (ANSI)
  - Process complies with WTO principles: Annex 4 of WTO/TBT Agreement
  - <u>All</u> stakeholders involved (Public & Private Sector Cooperation)
  - Neutral forum
  - Consensus-based procedures
- Development and delivery of information made uncomplicated
- A common sense approach: industry driven
- Market relevant globally
- No project costs

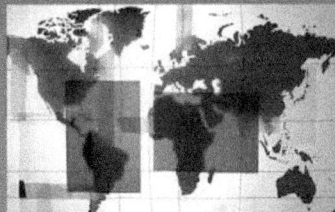

Presentation Credit: David Ober

E57 on 3D Imaging Systems

- Organized June 7, 2006 by Industry
  - Two Organized Meetings since E57 creation. (This does not include individual subcommittee meetings, task groups, or work items)
- Current E57 Roster: 113 Individuals and Organizations

E57 Committee Officers

- Chairman: Alan Lytle, NIST
- Vice-Chairman: Alan Aindow, Leica Hds
- Recording Secretary: Steve Hand, Survice
- Membership Secretary: Tom Greaves, Spar Point Research
- Member at Large: Tad Fry, Anheuser-Busch Incorporated
- Staff Manager: Pat Picariello, ASTM

E57 Scope

The development of standards for 3D imaging systems, which include, but are not limited to laser scanners (also known as LADAR or laser radars) and optical range cameras (also known as flash LADAR or 3D range camera).

The initial focus will be on standards for 3D imaging systems specification and performance evaluation for applications including, but not limited to:

. Construction and Maintenance
. Surveying
. Mapping and Terrain Characterization
. Manufacturing (e.g., aerospace, shipbuilding, etc.)
. Transportation
. Mining
. Mobility
. Historic preservation
. Forensics

The work of this Committee will be coordinated with other ASTM Committees and outside organizations mutual interest.

E57 Subcommittees

- E57.01: Terminology
- E57.02: Test Methods
- E57.03: Best Practices
- E57.04: Interoperability

E57.01 Terminology

- Scope: The Development of terminology commonly used for 3D imaging systems. The work of this subcommittee will be conducted with other ASTM E57 Subcommittees.
- Chairman: Gerry Cheok, NIST
- Vice Chairman: Kam Saidi, NIST

E57.01 Terminology – Update

- January 2007 – Approved ASTM E2544, includes
  - 8 terms specific to 3D imaging systems
  - Other commonly used metrology terms as defined by other standards
    - Accuracy
    - Bias
    - Calibration
    - Compensation
    - Conventional true value
    - Error of measurement
    - Indicating (measuring) instrument
    - Limiting conditions
    - Maximum permissible error
    - Measurand
    - Precision
    - Random error
    - Rated conditions
    - Relative error
    - Repeatability
    - Reproducibility
    - Systematic error
    - True value
    - Uncertainty of measurement

- May 2007 – Second ballot for additional terms to ASTM 2544
  - 15 terms submitted
  - 8 terms approved
  - Resolved most of negative votes at June 2007 meeting

E57.01 Terminology – Approved Terms

3D imaging systems
Angular increment
Beam propagation ratio
Beam width
First return
Flash LADAR
Instrument origin
Last return

Multiple returns
Pixel
Point
Point cloud
Second order moments
Simple astigmatic beam
Voxel

Terms means that the committee has agreed that these terms shall be defined.

E57.01 Terminology
Negatives to be Resolved and Re-balloted

- **3D imaging systems**
- Beam diameter
- Beam divergence angles
- Control points
- Registration
- Stigmatic beam
- Spot size

E57.01 Terminology
Subset of New Terms to be defined

- **3D image**
- Ambiguity interval
- Angular, lateral, range/depth, spatial resolution
- Field of view, Field of regard, Instantaneous field of view
- Interim tests
- LIDAR, LADAR
- Mixed returns
- Modulation transfer function
- Pixel cross talk
- **Range noise, error, bias**
- Registration error
- **Scan density / point spacing**
- Scan rate / frame rate
- **Different types of systems (e.g., TOF, phase-based, triangulation, pattern projection, structured light, Moire)**

E57.01 - Future Tasks

- Continue work on approximately 40 new terms
- Teleconferences every 2 weeks
- Contact: Gerry Cheok, NIST cheok@nist.gov

E57.02 Test Methods

- Scope: The development of standard protocols that will be used to characterize 3D Imaging System performance.
- Chairman: David Ober, Metris
- Vice Chairman: Darin Ingimarson, QUANTAPOINT
- Secretary: Mike Garvey, M7 Technologies

E57.02 Test Methods – Overview

Each Test Method: Define purpose of test
- Data Collection
    - Requirements: Environment stability, lighting, etc.
    - Setup General: Hardware (sphere, plane, reflectance) height, IA, Range, etc.
    - Setup Instrument specifics: Point Spacing, Dwell time, Data Rate, Internal filter settings, etc.
    - Measurements: Scan data (XYZ or RAE, RGB, Signal integrity: SNR or Intensity, etc.), temperature, pressure, humidity, light, wind, etc.)
- Data Analysis
    - Data Filters (Allow post processing manufacture filters vs. raw data)
    - Conversion Interoperability (data format)
    - Algorithms (Process the data)
        - Outliers vs. statistically meaningful data
- Results Report
    - Manufacture specifications (and how they integrate with the analysis AND results)
    - Data presentation (MPE, STD, Histogram, Mean, RMS, % Outliers, % data missing, % coverage, etc.)

E57.02 Test Methods – Update

Concentration: Scanners with Maximum Test Range < 120 meters

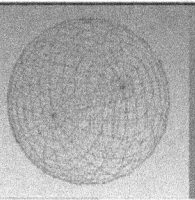

- **Range Uncertainty Protocol**: (Included Data Collection, Analysis, & Report) Tested at M7 Technologies & NIST.
  Protocol undergoing next revision.

- **Angle Uncertainty Protocol**: Two *data collection* approaches tested: Spheres and Flat Planes at M7 Technologies & Quantapoint respectively.
  (Analysis & report still TBD)

- **Planar & Spherical Analysis Simulation**: Analysis of existing Plane & Sphere Fit routines on detecting ***unbiased*** instrument Range, Azimuth, & Elevation uncertainty

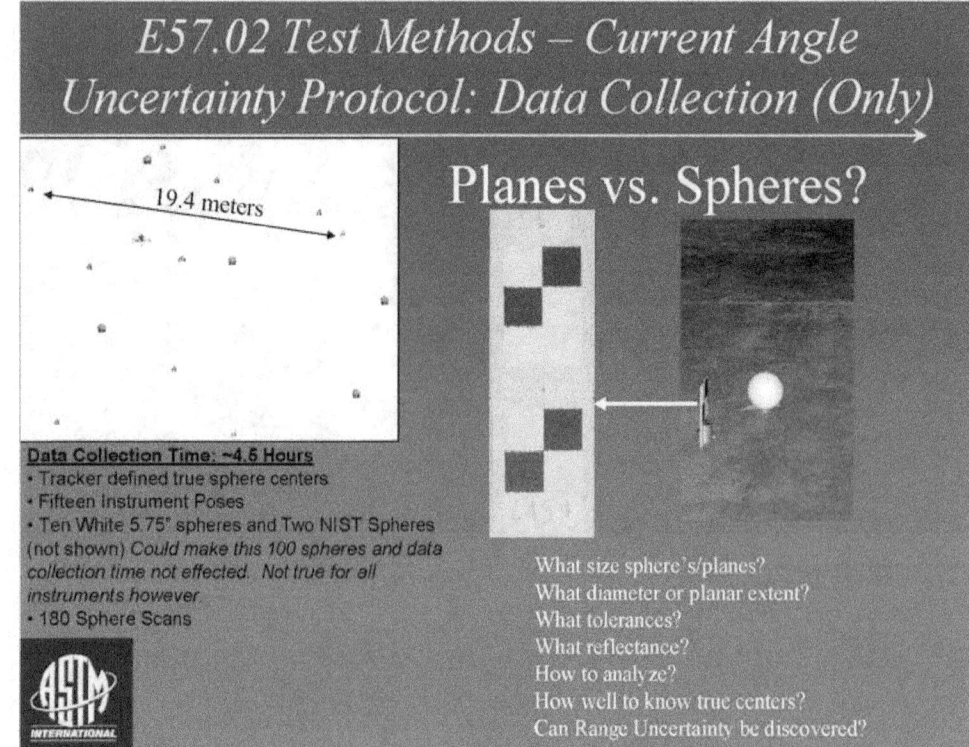

E57.02 Test Methods – Current Angle Uncertainty Protocol: Data Analysis (Only)

Horizontal

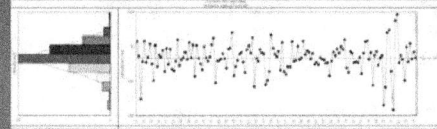

Results are presented in a compact format:
- Horizontal Angle 60.4 arc seconds
- Vertical Angle 47.0 arc seconds
- Distance 2.54 [Millimeters]
150 measurements

Vertical

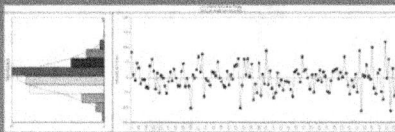

Distance

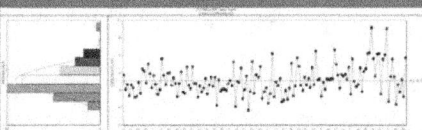

(These results were obtained for demonstration purposes ONLY and are not meant to represent an accurate analysis of any particular vendor's hardware)

Sphere's are not 5.75" (Tracker measures them ~5.735")
Analysis of calculated sphere centers not validated
What about Triangulation Scanners in this Spherical Frame (3D uncertainty only)?

E57.02 Test Methods - Simulation

INPUTS:
- Simulated detailed Range Noise model that included Reflectance, cos(Incident Angle), and Distance
- No errors in Azimuth and Elevation measurements. (measurements = true)
- Varied Scanner placement to plane and sphere (so data was not always collected at same locations on targets)

ANALYSIS:
- Minimize perpendicular distance of measurements to best fit plane/sphere.
- (Spheres) Constrained Fit to true sphere radius
- Sphere fit also used the covariance matrix of Range, Az, El to X, Y, Z to weight each measurement point in the fit.

RESULTS:
- Angle uncertainty bias is injected when using these Minimization Functions.
  (Sphere result)
  o Weighted = true (Un-weighted was worse)
  o Range = 10 meters, Point Count = 6774
  o RMS Range Measurements = 0.07 mm (~7 urad at 10m)
  o RMS AZ and EL Measurements = ~3.8 urad

Range RMS is reduced to 80% of total RMS value
Angle RMS is injected with 43% of total RMS value

E57.02 Test Methods – Up Next

- Meet Bi-weekly (or monthly as work progresses)
- Examine ISO/TC Terrestrial Laser Scanners Protocol: UNIBwM – 85577
- Complete the Range Uncertainty Data Collection Protocol
- Continue Analysis Simulation Study
  - Develop best fit routines to reduce bias transfer between 3 dimensions for planes and spheres. (NIST has an algorithm being developed)
  - Committee can decide to live with this bias transfer
- Develop the Range Uncertainty Analysis Protocol
- Develop the Range Uncertainty Results Report
- Test the new Range Uncertainty Protocol
  - Leica and Faro have indicated that they are willing to run these tests at their facilities.
  - NIST & M7 Technologies continue to also provide their facilities for testing as well.
- Future Tasks
  - Resolution uncertainty
  - Dynamic Range
  - Adapt tests to other instrument technologies (line scanners, Airborne scanners, etc.)

Contact: David Ober, david.ober@metris.com

E57.03 Best Practices

- Scope (Draft): Develop, validate, document and communicate best practices in the successful and consistent application of 3D imaging technology. Using these practices and guidance, end users can specify application requirements and associated deliverables traceable to accepted standards. Practitioners can determine instrumentation, procedures, and quality control processes yielding work product suited to application requirements.
- Co-Chairman: Ted Knaak, Riegl
- Co-Chairman: John Palmateer, Boeing

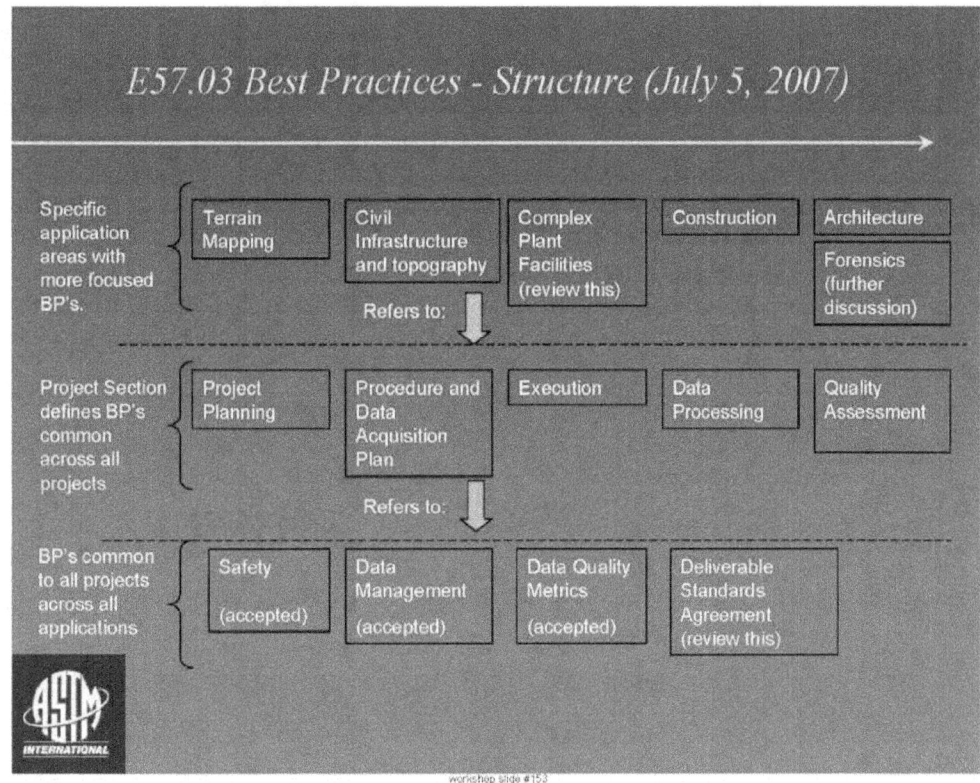

E57.03 Best Practices – Future work

- Vote on Scope Statement
- Develop Document Management (MicroSoft Teamshare or ASTM tool)
- Develop General Best Practice Outline (Introduction, Scope, Contents, etc.)
- Develop Task Groups:
  - Data Management Best Practice
  - Safety Best Practice
  - Data Quality Best Practice

Contact: Ted Knaak, tknaak@rieglusa.com

E57.04 Interoperability

- Scope: To develop and promulgate open, standard data exchange mechanisms for 3D imaging system derived data in order to promote its widest possible use.
- Chairman: Gene Roe, Autodesk
- Vice Chairman: Mark Klusza, Trimble

E57.04 Interoperability - Accomplishments to Date

- Held 3 meetings in 2007
- Developed and approved scope
- Established working committee
- In the process of defining each section of a draft requirements document

E57.04 Data Interoperability - Goals

- Research existing data exchange formats such as LAS and XML
- Seek the input and involvement of all interested parties
- Develop and proposed a Version 1, common data exchange format that meets the needs of the industry within 12 months

E57.04 Data Interoperability – 12 Month Work Plan

- Develop requirements definitions by 9/1/2007
- Draft requirements document by 12/1/2007
- Review and edit by 1/1/2008
- Deliver final to ASTM January 2008 meeting
- Develop draft of exchange format by 4/1/08
- Obtain approval by ASTM summer meeting 2008
- Interact with other standards organizations

Contact: Gene Roe, Gene.Roe@autodesk.com

E57 on 3D Imaging Systems

- We need you (manufacturers and users) to come help E57 define a successful 3D Imaging standard.
- Questions?

3D Vision, Robots and Movement – Putting it all Together

Brian McMorris

Robotics Industry Manager

Presentation Outline

- What is a Smart Camera?
- What is 3D?
- How can 3D imaging and Smart Camera technology be combined?
- Describe 3D image acquisition
- Application examples
- Describe some 3D tools
- Q & A

What is a Smart Camera?

- **Smart Camera:** "an integrated, intelligent camera combining all the "blocks" of a vision system (the image processor, camera and optics) within one device. It acts as a general purpose machine vision solution and ..." 2002

2D Smart vendors
- SICK|IVP
- Cognex
- DVT
- PPT
- Sharp
- IPD
- Keyence
- Omron
- Banner

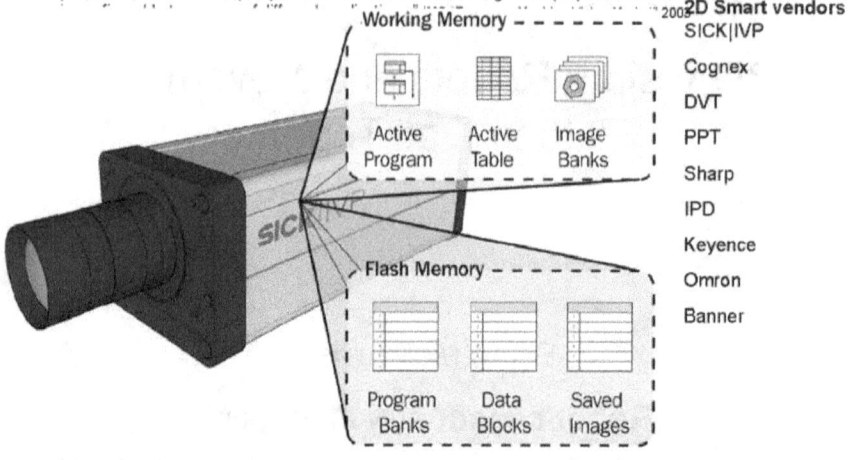

Applications Difficult to Solve with 2D

Missing objects

Low contrast applications

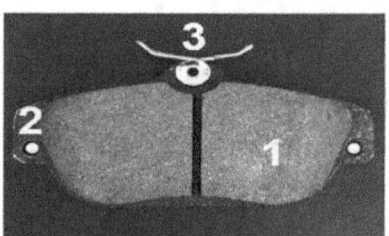

Bin picking random placement. X-X-Y-Z data required to pick object.

Encoder or Constant Speed

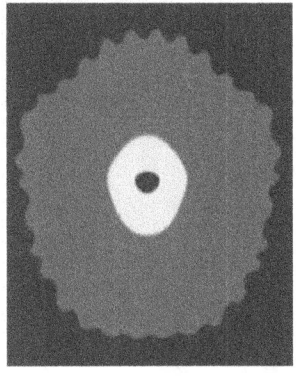
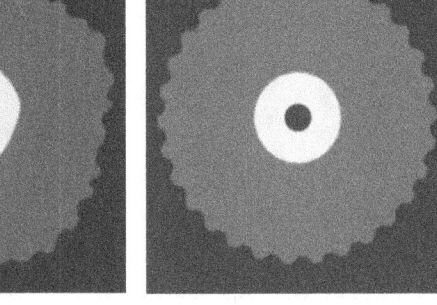
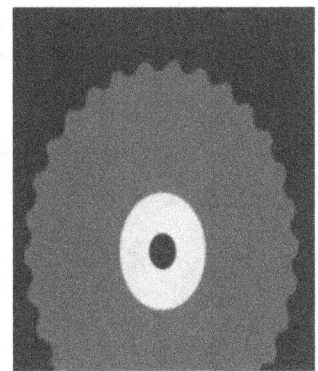

The velocity is uneven   Constant velocity or encoder feedback   The velocity is too slow

For robust implementations use encoders

3D + smart + tools

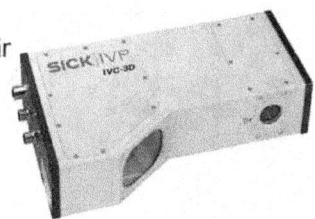

We combined a Smart Camera, tools and 3D imaging Camera!!!

- 3D imaging (3D image capture, encoders, calibrated units -> 3D image and profiles)
- Smart (stand alone, general purpose vision processor)
- Tools (well known 2D tools + 3D specific tools for profile and 3D image processing)

Smart 3D technology vs. Smart gauge

Smart 3D technology vs. Streaming 3D device

 Smart vs. Streaming technology(Multi sense)

Acquisition of 3D + intensity using a single laser. Processing too complex for current 3D Smart technology.

 Let's Solve a GOOD Application

Slice the Loaf Using 3D Vision Tools

- The number of slices depends on how hungry you are
- Volume of one slice = Volume of loaf/number of slices
- Set a thin ROI (one pixel high), moving ROI at the beginning of the loaf and set the accumulated volume to 0.

For number of slices:
- Calculate the volume of the thin part of the loaf that is inside the ROI and add this volume to the accumulated volume.
- Move the ROI one pixel at a time and add the volume inside the ROI to the accumulated volume.
- When the accumulated volume becomes larger than the desired volume of a slice, cut the loaf (or just mark the location) and reset the accumulated volume.

The Steps

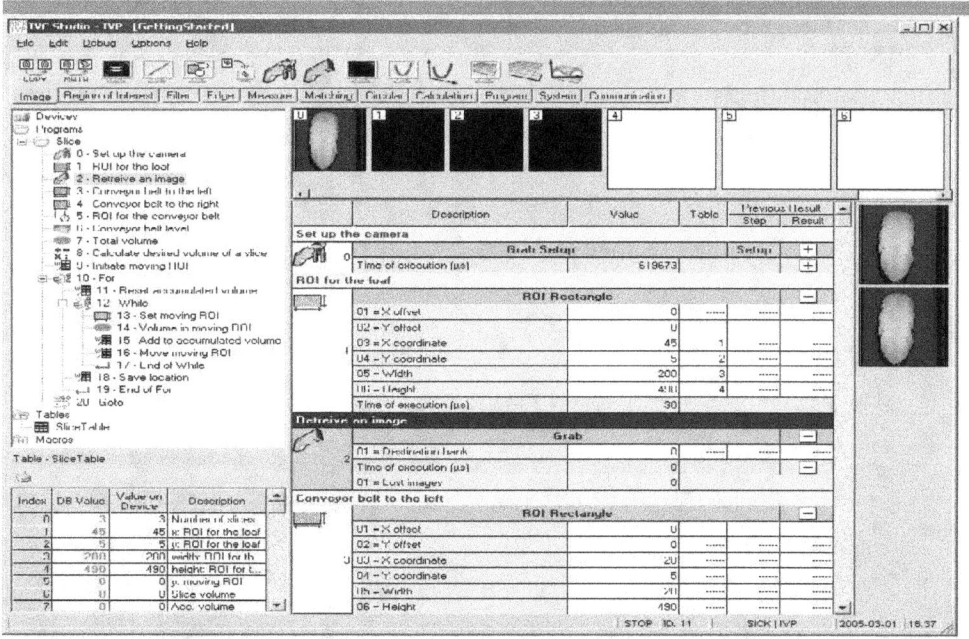

Dedicated 3D Image Processing Tools

Edge tools

Calculation tools

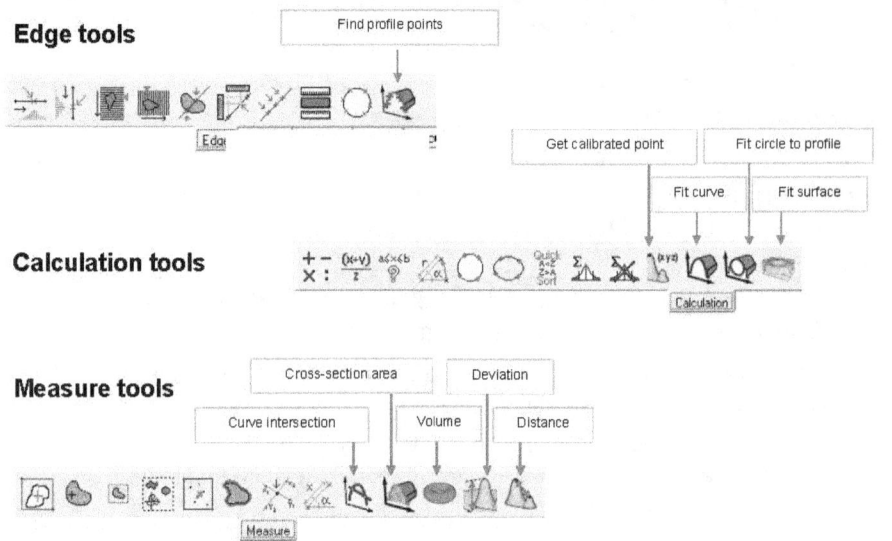

Measure tools

Dedicated 3D Image Processing Tools

Image tools

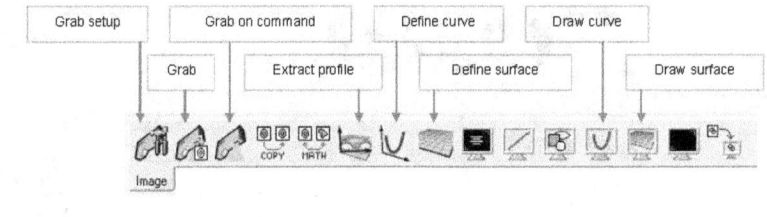

ROI tools **Filter tools**

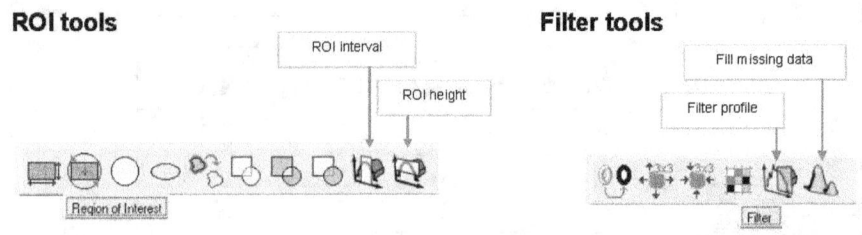

Brake Pad Inspection

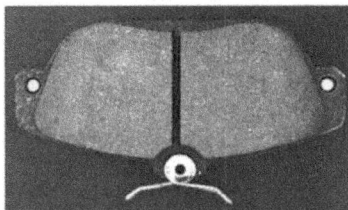

How is the spring angle and height measured with 3D tools?

- Brake Pad Demo
  - 0 - If Error Goto
  - 1 - Initialization
  - 2 - Wait for trigger to go high
  - 3 - Reset the first image bank
  - 4 - Reset Fail flag
  - 5 - Start an image aquisition
  - 6 - Reset all other image banks
  - 7 - Place the new image in bank0
  - 8 - Detect Surface Cavity
    - 9 - Define the entire image bank
    - 10 - Define the base surface
    - 11 - Draw the base surface
    - 12 - Add all pixels higher than the base surface
    - 13 - Find one point that will allways be on top of the object
    - 14 - Find the outline of the top of the object
    - 15 - Define a Region of Interest inside the top of the object
    - 16 - Fit a surface to the top of the object within that ROI
    - 17 - Find the height of that surface
    - 18 - Locate the surface cavity within the ROI
    - 19 - If then Goto
    - 20 - Add Text
    - 21 - Set Fail flag
  - 22 - Detecting cable
  - 29 - Jump on Detected Defects
  - 30 - Print Pass Text
  - 31 - Display
  - 32 - Read Input

Industrial Configurations for 3D Smart

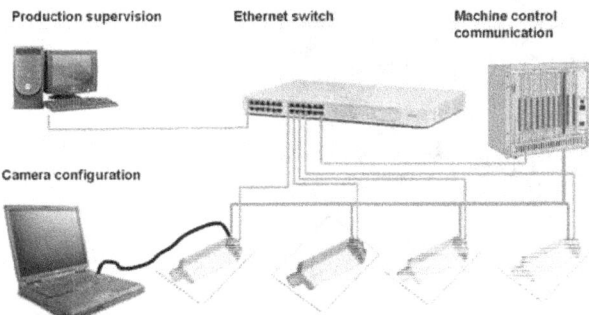

- 3D and 2D Smart can be used in a variety of configurations:
  - Stand-alone single camera unit
  - Stand-alone multiple camera unit
  - Managed by a control system
  - Monitored by a PC

3D Gray Scale = Z height

A 3D image shows the topology of an object, or the distance from the bottom (or reference plane) to a point on the surface of the object. The brighter a pixel is in the image, the higher up that point is on the object.

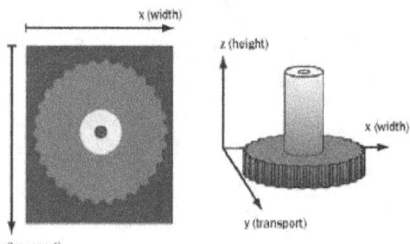

Finding the Volume

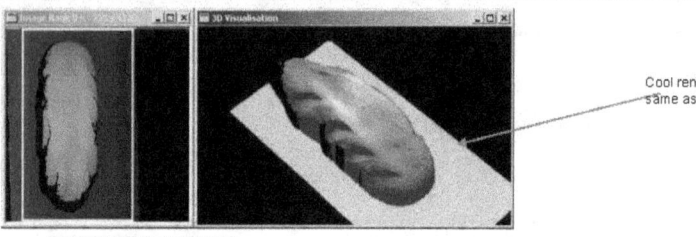

Cool rendering information is the same as other image

- **ROI Rectangle (find the bread)**
  - The rectangle that is the region of interest (ROI) in which we expect to find the loaf or the conveyor belt.
- **ROI Union (find the conveyer)**
  - Two ROIs join. An ROI is specified by referring to the program step in which the ROI is created.
- **Fit Surface (find the crust)**
  - An image bank containing a 3D image, an ROI, and the type of surface to fit to the part of the 3D image inside the ROI.
- **Volume (calculate the bread volume)**
  - An image bank containing a 3D image, an ROI and a surface to use as zero-level when calculating the volume inside the ROI. Anything in the image below the zero-level is ignored when calculating the volume.

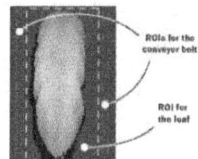

ROIs for the conveyor belt

ROI for the loaf

IVC-3D Smart Applications

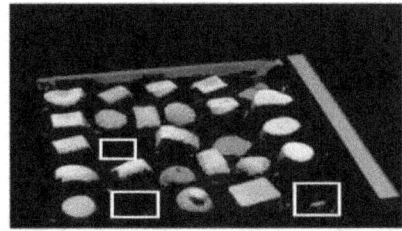
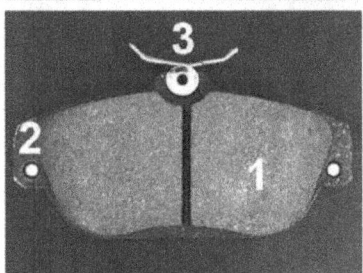
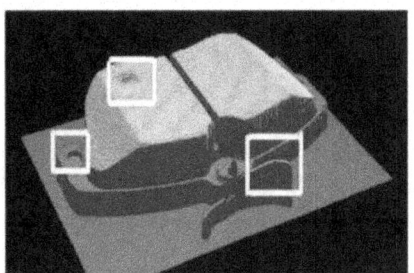

Ideal 3D Applications for SICK Vision

- Weld Seam and Glue Bead Inspection: Vision Guided Robotics
- Part Picking: Random Orientation in Bin, Auto Racking, Conveyor, Chain Hangar
  - Machined and cast parts with non-square edges (poor shadowing)
  - Washers (not flat), extrusions, metal stamping (no contrast)
- Palletizing, Depalletizing, Stacking, Case Packing
- Machine Tending: Load and Unload : Vision Guided Robotics
- Low contrast imaging applications: rubber, plastic, asphalt, baked goods
- All types of volume surface feature applications:
  - Metal machined and welded parts
  - High speed surface inspection, e.g. In-motion Railroad rails and rail beds, highway surface quality (mapping potholes and cracks)
  - Packaging (confirm integrity of boxes, presence/absence of product)
  - Baking and cookie inspection
  - Tires, gaskets, automotive trim parts (low contrast, non-squared edges)
  - Pharmaceutical applications (blister and fill levels)
  - Turbine blade inspections
- Size distribution by volumetric calculation (single and multiple objects)

Robot Auto Industry Applications

Welding

Handling

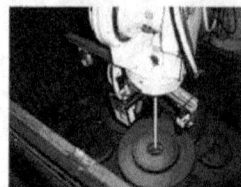
Machining

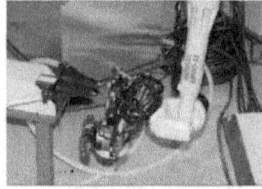
Assembly

Glue Dispensing

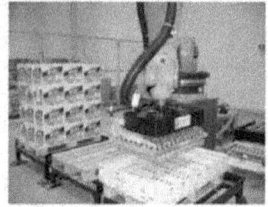
Picking / Palletizing

Common use of Robots and 2D Vision

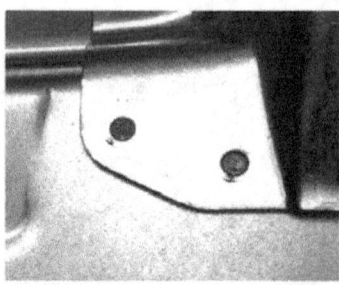

- Inspection for weld presence and position using 2D vision

Found welds = 0

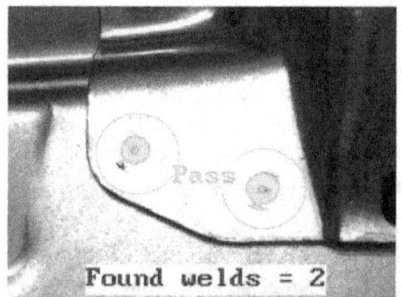
Found welds = 2

What is 3D Profiling?

Detection, Positioning and Inspection with 3D Shape

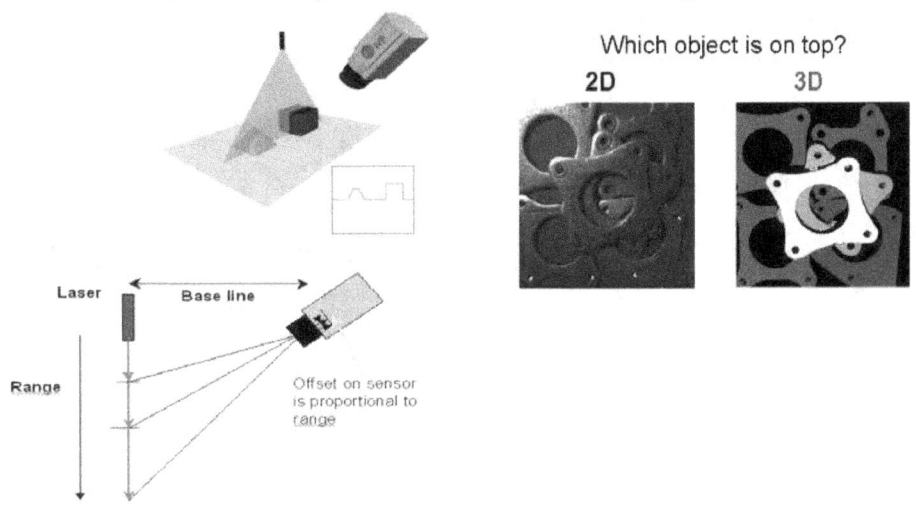

3D Vision Buzz Words

- Occlusion: Laser and lens
- Laser class: Class II eye safety. Can affect frame rate
- Resolution: (for guidance only)
  - Cross (X), Width/number of pixels
    - Example 500 mm wide belt, a sample every 0.5 mm
  - Movement direction (Y) Velocity/ frame rate
    - Example 1m/s/5000 -> 5 profile/ mm -> 500 profiles for 50 mm long object
  - Height (Z): Depends of FOV (Sub pixel techniques)
    - 1 inch ~ 5 microns (1 inch FOV (width) is possible to achieve about 5 microns Z resolution
    - 5 inches ~ 1/1000 inch
    - 30 inches ~ 1/100 inch

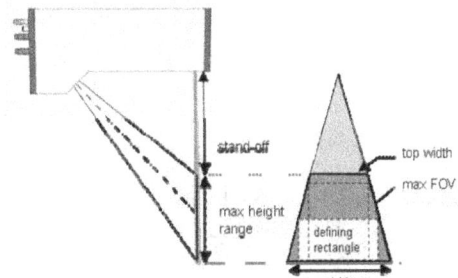

3D Calibration Basics

Camera FOV in real world

Floating point position (x, z)

Floating point position (u, v)

Sensor

Sensor points not positioned on pixel grid

Given:
- a set of point pairs (u, v) and (x, z)

Wanted:
- the value of x for each pixel (u_0, v_0)
- the value of z for each pixel (u_0, v_0)

Problem solved by interpolation.

Local Surface Fitting Model Calibration

Find the corresponding x0 and z0 value for any sensor pixel (u0,v0) by interpolating using a surface fitting model.

The surface fitting model should interpolate from the 6-10 nearest neighbours to get both the x and z value.

Generate 2 look-up tables. One for x and one for z values for each sensor pixel.

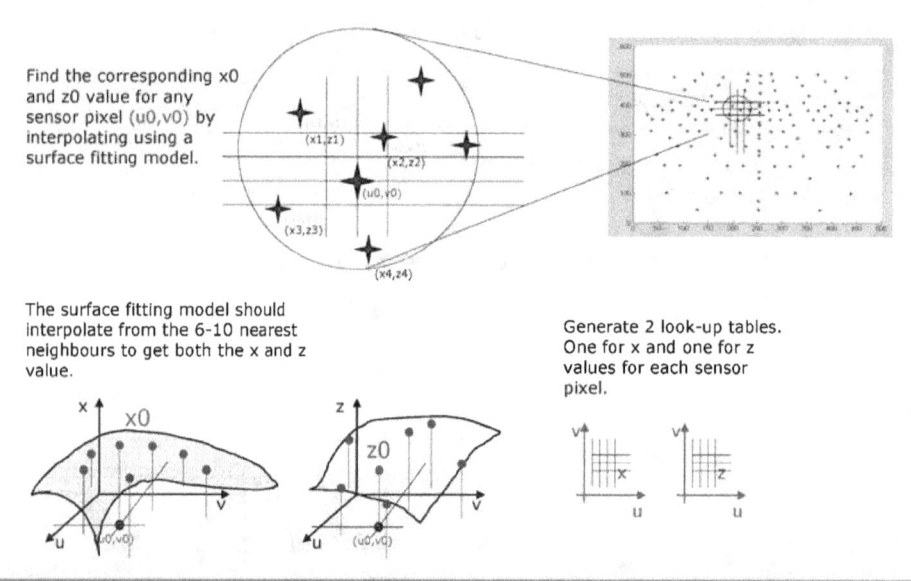

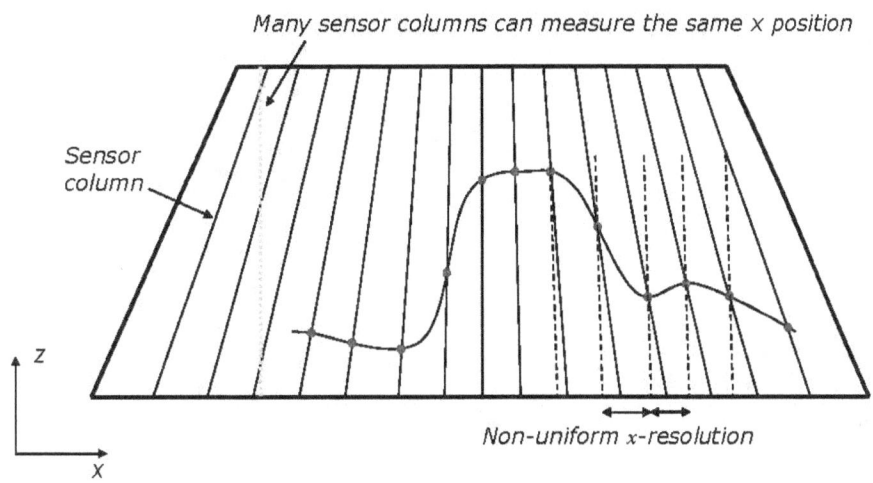
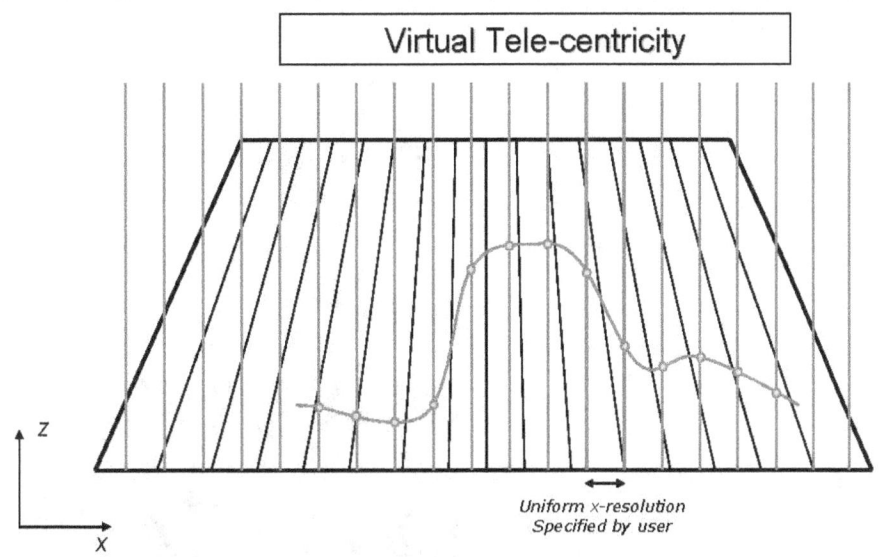

 ## Delta Robot Bread Packaging

- A 3D camera locates bread buns for packaging by a delta robot
- Faulty buns are rejected
- Stand-alone operation, no PC needed for image processing

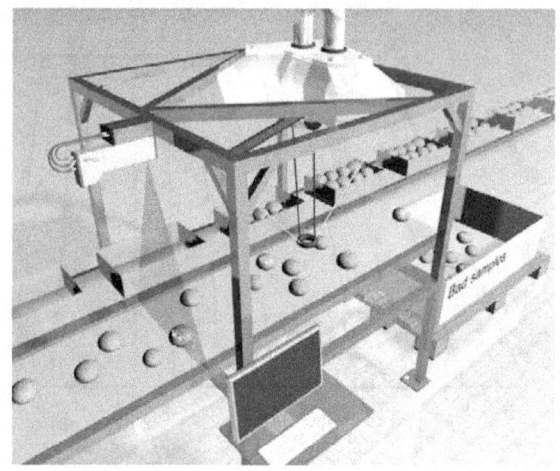

Industrial Sensors • Safety Systems • Automatic Identification Solutions

 ## Camera Robot Calibration

- The 3D camera delivers results in robot coordinates!
- Calibration method
    - Mount the camera
    - Scan a calibration object. The camera measures key positions
    - Point out the key positions with the robot
    - The camera calculates the coordinate transformation

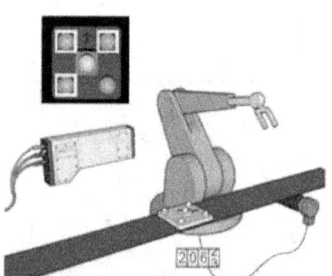

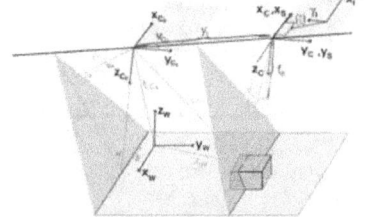

Calibration object

Industrial Sensors • Safety Systems • Automatic Identification Solutions

Conveyer and Robot Calibration

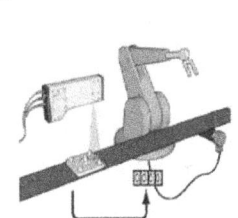
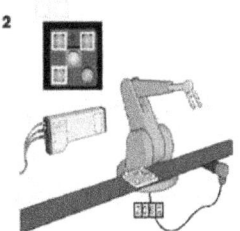
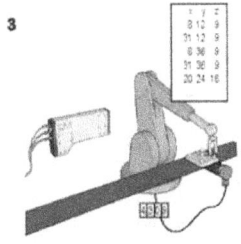
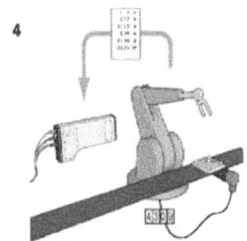

1. Grab a 3D image of the reference object. The Smart Camera sends a trigger pulse to the robot

2. Mark the reference points in the 3D image and move the object to the robot. The robot uses an encoder to keep track of the object's movement.

3. Get the robot's coordinates for the reference points.

4. Import the robot's coordinates – if necessary, adjusted by the movement – to the Smart Camera and calculate the transform.

Industrial Sensors • Safety Systems • Automatic Identification Solutions

Benefits of using 3D Vision in Bin Picking

- Allows picking of complex products
  - 3D shape is often much more important than 2D pattern when picking up objects
  - Does not require unique features for part location
- Contrast-independent inspections
  - Dark products on dark conveyor
  - Color-insensitive
  - Insensitive to dirt or patterns
  - Robots often handle products before their appearance has been finalized (e.g. painting)

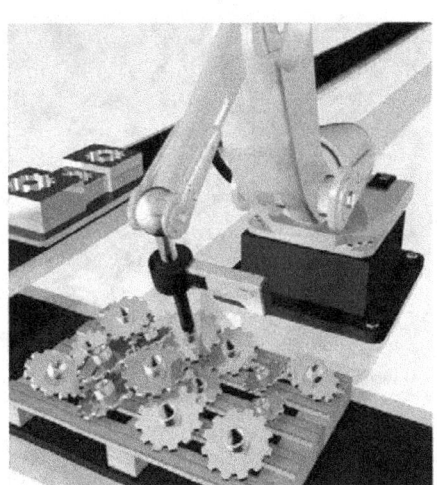

Industrial Sensors • Safety Systems • Automatic Identification Solutions

Bin Picking

Some of the factors that determine bin picking complexity

- Part: Shape, surface features, material, size, fragility

- Presentation: Bin with/without sides, random, matrix, stacked parts, layers

- Required throughput (parts/second): Robot movement capabilities, part scanning

- Precision: System requirements for pick point precision and the ramifications for calibration, transformation and robot position

- Other: Collision avoidance requirements, robot interface

3D Robot Calibration, similar to 2D

Calibrating robot with SICK 3D reference tool

Transforming vision system coordinates to robot coordinates

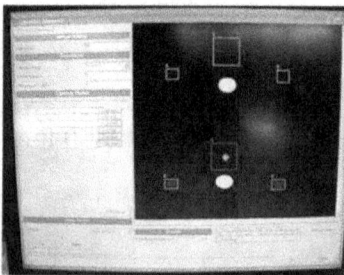

 # SICK Bin picking at IRVS '07

Example of random bin picking using Smart Camera (IVC-3D-200); no fixtures and non-uniform objects to pick

 # Random Bin Picking is Here

- Acquire 3D image of objects
- Report coordinates and orientation in 6 DOF (Degrees of Freedom) to the robot controller
- Robot picks the approriate object

Fraunhofer Institute IPA & RoboMan (using IVC Ranger camera)

Master Automation

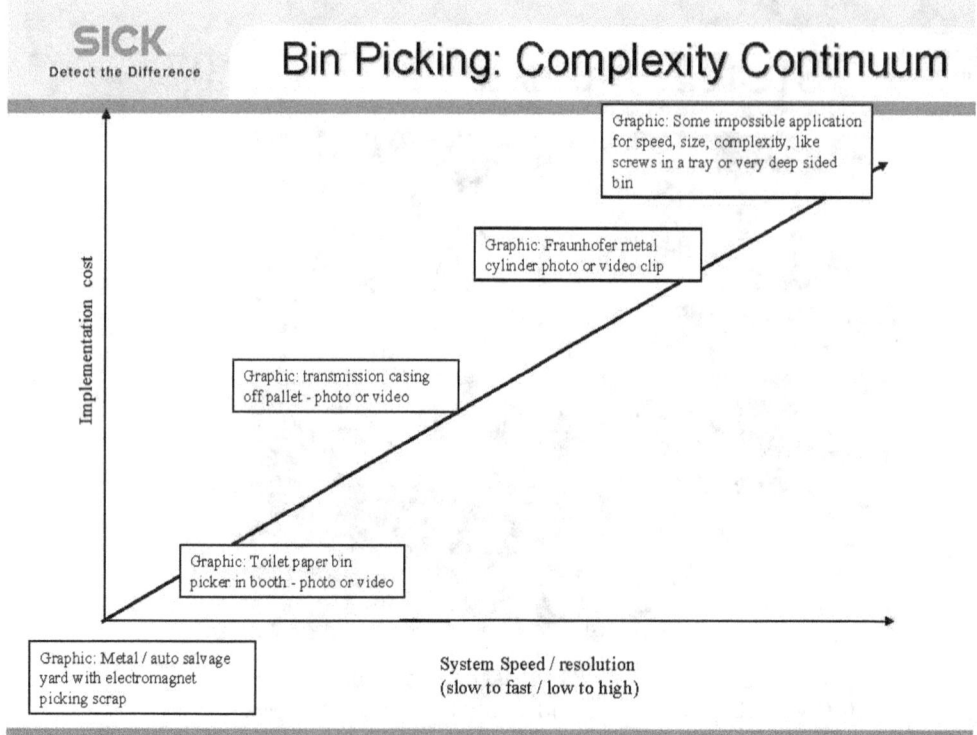

Painting of Car Bodies: VGR

- The 3D shape of the car body is measured and reported to the robot controller
- Optimal paint-stroke pattern is calculated
- Painting starts

Vision-Guided Robots: Palletizing in Packaging

- A 3D system is used for contrast-independent palletizing
- Packages are located on the conveyor, then placed correctly on pallets

Crate Handling

- A 3D camera scans the crate and locates all bottle caps
- Contrast-

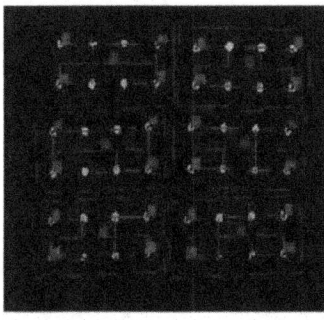

Depalletizing in Automotive

- A 3D camera is used for exact location of a gear box part on a pallet

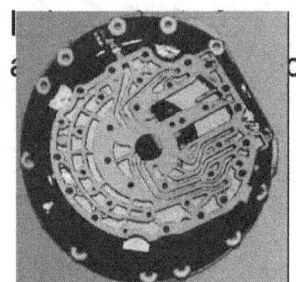

Glue Dispensing/Weld Seam

Direct robot control from 3D smart cameras (no PC or remote controller needed)

• Monitor volume, shape, height and width of weld/glue

•Feedback for dosage control and weld parameters

•Smart 3D device will control robot directly via ethernet/serial

•Robot will take appropriate action, such as increase glue volume, reapply glue, increase weld temperature, etc.

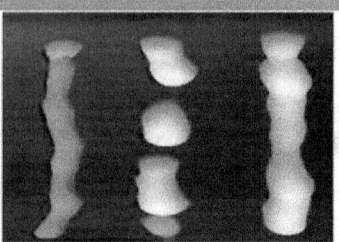

Glue beads

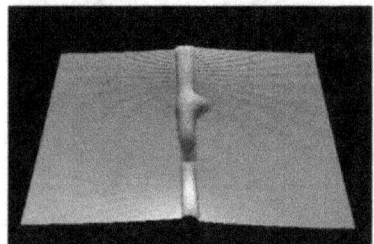

Weld seam

3D Images of Chocolates

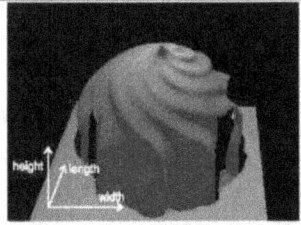

3D image of chocolate

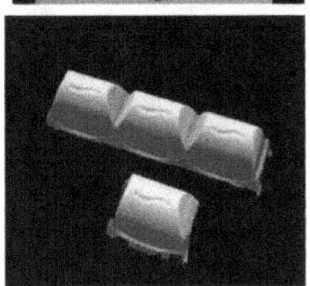

More chocolate

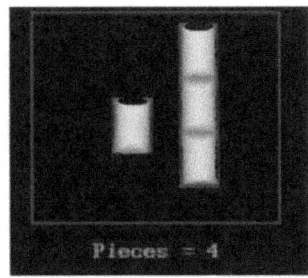

3D Smart Output-> OK product. Four pieces of perfectly shaped chocolate

De-palletizing

Handling non-contrast items

Handling various contrast items

Courtesy of KUKA

Image Recognition

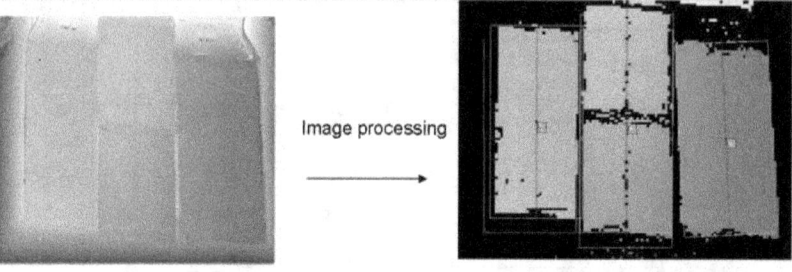

Image processing

Filtering combinations of erosion and dilation (opening and closing)
Blob search with adaptive 3D height gray values
- 3D gray values limits based on histogram data
- Found blobs (connected areas of specific 3D gray height value) are evaluated regarding their length and width to match the searched articles

Blob matching
- Unidentified blobs are reevaluated to match double sized packages

Industrial Sensors • Safety Systems • Automatic Identification Solutions

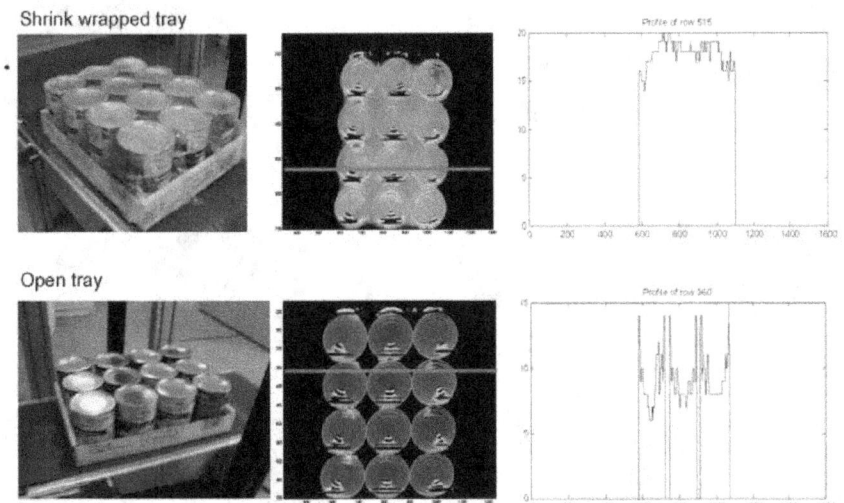

Company & Technology Overview

Gaile Gordon, VP of Advanced Development

WWW.TYZX.COM

TYZX Overview

- Moving the physical world online
  - Systems that *see, interpret and respond* to the real world
  - Real time Information and control
  - Commodity 3D perception systems for smart products
- Applications
  - Interactivity with users, real world environments
  - Automation, mobile robot navigation and safety
  - Vehicles, security systems, defense systems, entertainment
- Approach
  - 3D passive optical sensing
    - Provides 3D data and registered 2D appearance image
  - Common hardware/software platform for multiple applications
    - Cross market technology synergy
  - Custom ASICs for performance and cost
  - "3D Middleware" and development platforms to reduce OEM's TTM

Stereo Vision

- → Dense stereo vision
    - Uses local texture to estimate depth for *every* pixel
    - Expensive operations need custom hardware
- → Benefits
    - Full frame of 3D data at high frame rates
    - Operation in ambient light (passive)
    - Works with a variety of sensors (IR, UV, color, ...)
    - Flexible operating range through choice of baselines/lenses
- → Use where
    - Speed & latency are important
    - Environment is poorly constrained (natural scenes and objects)

Markets

- → Automotive
    - Pedestrian detection
    - Lane keeping
    - Occupant sensing
- → (Person) Tracking & counting
    - Security and Surveillance
    - Entertainment and Marketing
- → Robots
    - Autonomous systems
        - Defense and consumer
    - Vision guided vehicles
    - Vision guided servoing
    - Safety applications

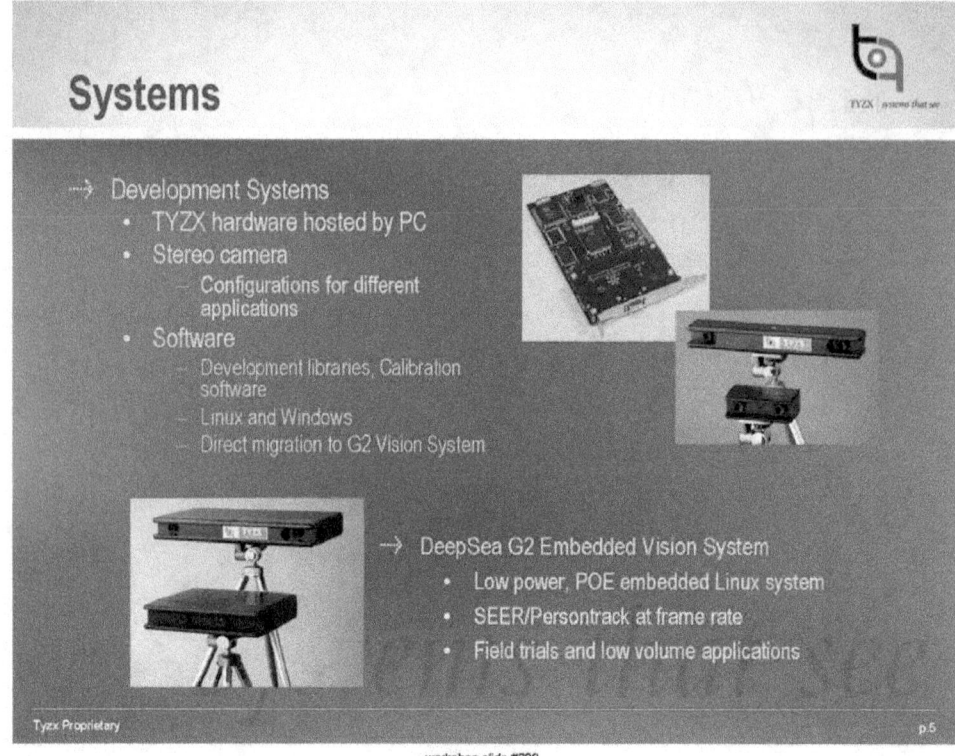

workshop slide #209

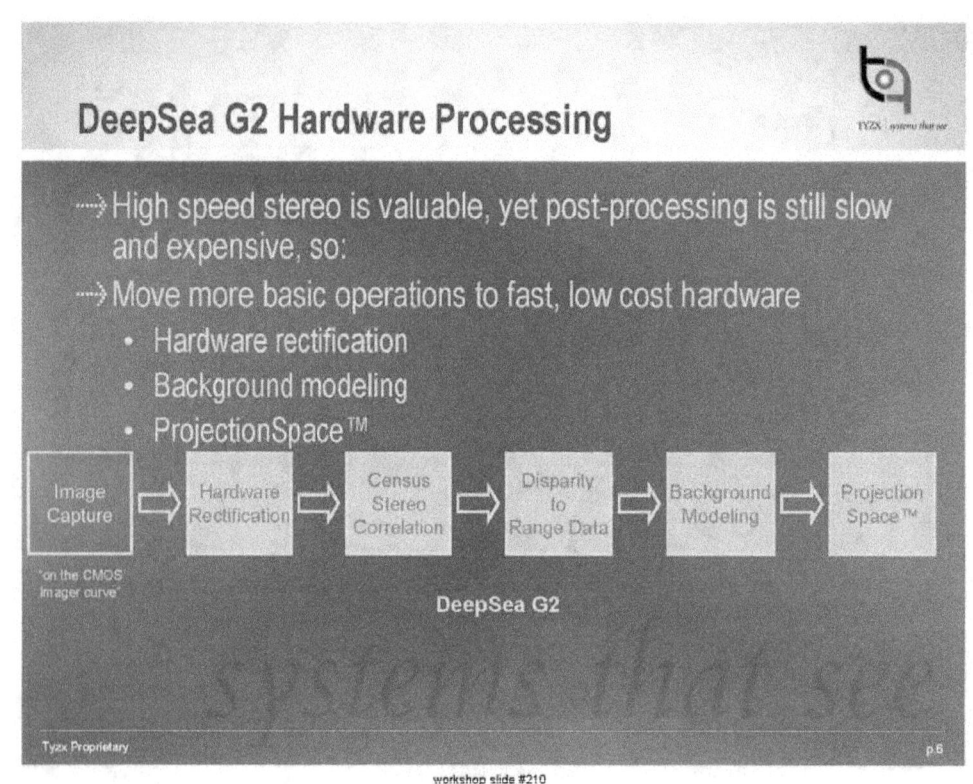

workshop slide #210

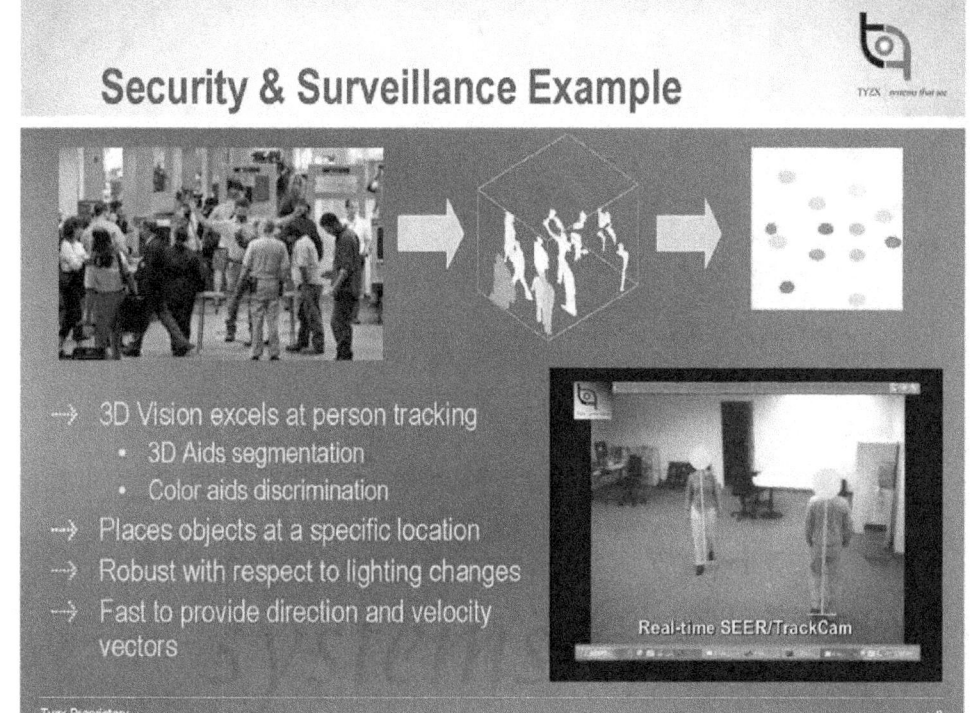

Robust, Precision Stereo Cameras

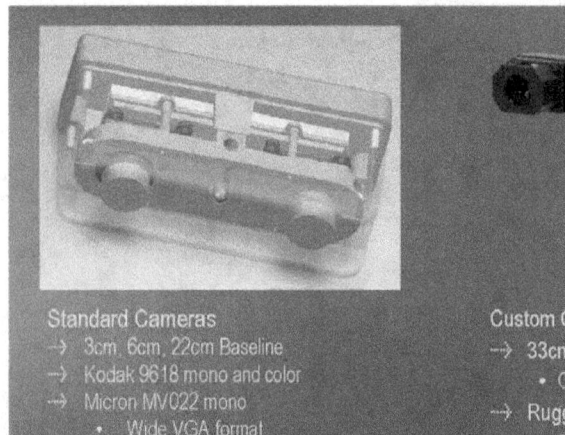

Standard Cameras
- 3cm, 6cm, 22cm Baseline
- Kodak 9618 mono and color
- Micron MV022 mono
  - Wide VGA format
  - Full frame shutter
  - HDR
  - Color
  - 60 fps
- Robust construction

Custom Cameras
- 33cm baseline
  - Custom baselines available
- Rugged
  - Stiff, aluminum construction
  - C-mount with "Lens Cage"
- Exposure control
  - Unified motor iris, integration time & gain
  - HW assist

Quick Survey

- Stereo vs. 2D video
  - More reliable and easier segmentation and application processing
- Dense stereo vs. Feature-based and multi-camera
  - Better for capturing unplanned events
  - Imager pair maintains calibration/registration
  - Less suited for photogrammetry, though std triangulation can still be performed
- Stereo vs. range sensors (radar, lidar, sonar)
  - Passive
  - Higher resolution
  - Faster, less noisy
  - Wider FOV, flexibility with respect to operating distances
  - Stereo vision provides "fused" appearance image for free
- One technology – many sensors
  - Just change stereo baseline & lens
  - Millimeters to 100 meters

workshop slide #215

How to Implement Bin Picking in your Manufacturing Operation

Adil Shafi
President
SHAFI, Inc.

workshop slide #216

Abstract
This article is targeted towards the End – User manufacturing community. It is intended to provide a brief overview of Bin Picking's progress towards reliable and widespread use, with Vision and/or Light Guided Robotic techniques, and then to provide a methodology to consider, carefully test, and implement reliable Bin Picking.

Turning the Completely Impossible into the Obviously Possible
When Thomas Alva Edison began to work on inventing a light bulb, it was generally considered an unreliable and impossible task. When with self belief and perseverance he succeeded, he looked back and said that he had to succeed since he ran out of methods that could not succeed. Today, satellite images show impressive images of lights in industrialized regions on earth at night.

Learning from failures and the experience of others before him, Sir Edmund Hillary defied the conventional reservations of his time and summitted Mount Everest. Today, so many people summit Mount Everest each year that it is commonly joked that soon we will have a weather insulated escalator to go up to the top.

In our manufacturing community we have similar parallels. A generation ago, most welding was done by people, often with inspectors after welding stations. Today, manual welding is questioned and rare. Just six years ago, 3D Vision Guided Robotics performing AutoRacking (or pick or place stamped metal parts from or onto racks) was virtually unprecedented. Presently, we have hundreds of cells running AutoRacking reliably in our industry and some companies implement AutoRacking on every new manufacturing program.

I believe that the same is true of Bin Picking. A few solutions have been running in production for more than three years and more are being implemented each year. Within a decade or so, all Bin Picking will be automated. Our next generation will wonder why people would want to pick parts manually, more slowly and more expensively than a fast robot from a bin. Manual bin picking will then become questioned and rare.

The Enablers
Bin Picking, in the past three years, has quietly but steadily made advances in commercial production lines. A good review of successful solutions in our manufacturing industry was published in Automation World's February 2006 issue, www.automationworld.com. The article was entitled "Vision Guided Robotics: In Search of the Holy Grail".

Page 2

workshop slide #217

Ease of Bin Picking is driven and prioritized by two factors: 1) The geometry of the part, and 2) The degree or severity of randomness of parts in bins. The first, easiest, and financially most justifiable solutions have been in the automotive powertrain area; most notably engine blocks. These parts are well machined, are rich in geometric features, skewed slightly in x, y, z, yaw, pitch, roll directions and are heavy (thereby slow and hence expensive to manually handle). This has been a perfect first storm to enable Bin Picking.

There are many enablers currently driving more solutions into the fold of reliable Bin Picking. These include: Advances in computational processing power, vision recognition tools, mathematical algorithms, flexible lighting, a continuous reduction in commercial pricing, and a growing collection of techniques in handling, gripping and staging an overall problem into more easily handled steps.

A rough analogy is that 16 = 4 x 4, but 16 is also 4 + 4 + 4 + 4. Addition is easier to do than multiplication. The same problem can be reduced into several smaller equivalent problems.

A tough bin picking challenge can be simplified by breaking the problem into individual retrieval only first, which may be imprecise in finding a part centroid, but then using a simpler 2D pick and place stage for precise final placement. Such two-stage operations can reliably run entire bins and meet a six second part-to-part, bin acquisition to precision pins placement cycle time. Fast, fixed mount camera solutions are now running in production at four second part-to-part cycle times.

Good Applications That Are Ready for Reliable Bin Picking in Production Now
The following applications have now become feasible for reliable Bin Picking:

1. **Automotive**
    - PowerTrain (Engines, Cylinder Heads, Axle Shafts, Differential Carriers, Pinions, Round Parts with Stems, Connector Rods, Piston Heads, Brake Rotors and Stacks of Gears).
    - Stamping (Flat or bent metal plates with multiple holes, roughly stacked stampings with a progressive skew).
    - Final Assembly Products in Boxes in T/C/F (Trim Chassis Final) pick operations for placement into cars on moving lines; see related discussion about Vision Servoing at the Robotic Industries Association website http://www.roboticsonline.com/

workshop slide #218

2. Packaging
- Strips of medical tablets, flat but randomly arranged in boxes.
- Bags of products e.g., chips, salsa, cheese, cement, etc.
- Lateral or upright layers of tubes (copper, plastic, PVC).
- Layers of products e.g., wooden planks, plastic sheets.

How to Implement Bin Picking in your Manufacturing Operation

The following steps are recommended to evaluate, justify and implement Bin Picking.

The instructions below are a bit precise but not difficult to follow.

1. Take pictures of your parts with a cell phone or a digital camera from an electronics store. You will need two cameras for your part and bin image analysis.

Individual Part Pictures (IPP)

2. Consider each part that you manufacture. Place each of your parts on a flat surface. Review the multiple stable resting positions in which each part can be placed on a flat surface (for example, a soft drink can has two stable resting positions: One "standing up" with its circular footprint on the flat surface, and one "lying on its side" with its circular planar ends perpendicular to the flat surface (the resting position in which it can roll on a flat surface).
3. Then for each of your parts, take a picture of each Stable Resting Position (SRP). The camera should be aimed at about a 45 degree angle to the flat surface, looking down towards the part. Collect this as your library of Individual Part Pictures (IPP). This is essentially a two-dimensional array of pictures, where the first index is your part number, and the second index is the part's SRP.

Bin Randomness Pictures (BRP)

4. The next step is to take each of your part types, and review how randomly they are found in actual bins in your manufacturing operation.
5. Using a tripod or a temporary structure, setup two fixed-mount cameras above each bin. Depending on the size of your bin, adjust the size of the view so that the Field of View (FOV) of your image is indeed the entire bin. Place the first camera directly above the bin pointing straight down or perpendicular to the flat horizontal plane of the bin below. Let's call this Camera 1 or C1. Place the second camera at a 45 degree angle above the bin, looking downward, so that it sees the C1 scene from an angle from any side (select one fixed side) of the bin. Again setup the FOV so that it has as much part content in it as possible as what C1 can also see. Let's call this camera at 45 degrees Camera 2 or C2. When looking at a bin, the planar 2D views of C1 and C2 will not be in the same direction nor scale and the C2 images will be skewed and that is fine.

Page 4

workshop slide #219

6. Then for each of your parts, place a bin of parts below C1 and C2 (as they normally occur in production to the level or randomness that you typically find them). Take multiple pictures of each bin and several examples of randomness of parts that you will see. Organize and maintain a pair of C1 and C2 image pairs for every scene.

Take This Pictorial Information to the Experts: Evaluate and Believe by Seeing Demos

1. Take this pictorial information to experts in the field of Bin Picking. You can use an Internet search engine (enter "Bin Picking"). Request examples of their past work as well.

    You can also attend and meet speakers at the 3D Bin Picking Conference track at the International Robots & Vision Show in Rosemont, Illinois (Chicago) on June 12 – 14, 2007 http://www.robots-vision-show.info/robots_vision_show_info.html. There will be several Bin Picking demonstrations running at the show.
2. Request an evaluation of your parts from the pictorial information collected above. It is then possible to obtain a budgetary estimate to automate your Bin Picking operation. If the payback on investment is justifiable, then proceed with the following steps.
3. The first key to success is to insist on a pre – sale demonstration with exactly your parts. This is a critical step to not misunderstand and to not create failures. It is very important to ask for a completely reliable, uninterrupted retrieval of all parts, or negotiated manual intervention for certain cases of part randomness. It is the only way to adequately protect the risk in these projects for five parties : End – User, Systems Integrator, robot company, vision company, and software enabling company.

    Sometimes these roles are provided by the same company, however Bin Picking experience and a single line of project responsibility from a Systems Integrator is critical to your success in this area.

Seeing is Believing

It is highly recommended that your factory floor personnel visit and review vendor demonstrations, since they often know of rare and exceptional cases that can stop production. It is critical to gain a comfort level by seeing several, continuous, uninterrupted and realistic demos running from completely full to completely empty bins before issuing a purchase order.

Part Variation Management

4. The second key to success is Part Variation Management (PVM) in your operations. It is very important to separately study, log, plan and manage manual–to–automatic

Page 5

workshop slide #220

retrofits versus new part programs. In a retrofit situation, it is possible and recommended to take hundreds of unobtrusive pictures (see C1 and C2 image gathering methods above), and to be able to run simulated pickups of those images offline.

This process protects being caught off guard after good laboratory demos and runoffs at the vendor site, while remaining unaware of true variation in a plant. Sadly, this is often realized late in a project when the vendor arrives at the End – User plant for final implementation, only to discover that a number of variation cases were unexpected, misunderstood and unplanned for in advance.

These types of mistakes create disillusionment and delay in future confidence, and ultimately delay the time advantage in financial benefit to End – Users. It often takes a year or two for a typical End – User to recover, reinvestigate and reinvest. In the meantime, other global End – Users gain competitive advantage by avoiding these mistakes.

5. Thirdly, it is recommended that you review and benchmark, through actual test, ease of use for non – technical operators, training at Operator, Technician and Engineering levels, a FMEA (Failure Mode Engineering Analysis), and rigorous procedures for backups, version control, and access to 24/7 vendor support.

Conclusion

Bin Picking is a manufacturing solution whose time has now come. There are many examples of Bin Picking that are ripe for success and financial benefit to End – Users. The content above provides a methodology for analysis and evaluation. It also provides project management guidelines critical to protect End – User success.

workshop slide #221

The History and the State of the Art of Laser Tracker Technology and Applications

Kam Lau, Ph.D.
President
Automated Precision, Inc.
Maryland USA

Metromeet 2008

Topics of Discussion

- Brief introduction
- History of Laser Tracker Development
- Theory of Laser Tracking and comparisons of different tracking techniques
- Evolutions of laser tracker designs
- Tracker Traceability (ASME B89.4.19 Vs USMN)
- Advances in tracker applications and accessories
- Summary of Discussion

- Ph.D. in ME, U of Wisconsin –Madison 1982
- Worked for NIST (formerly NBS) between 82-87
- Founded Automated Precision, Inc. in 1987
- Major metrology system inventions:
    - 3-/6-D Laser Tracker (Co-inventor Dr. Robert Hocken)
    - I-probe, Iscan, 6-D Smart Track, RTOF ADM
    - RapidScan– a 3-D optical surface scanner
    - 3-D locking high-precision CMM probe
    - Deep Bore Profiler for measuring profile of deep holes
    - XD Laser Interferometer System for simultaneous multi-axis CMM/machine tool error measurements
    - CNC Machine G/T Error Model and Compensation

API Headquarter at Rockville, MD USA

ISO 17025 metrology lab accreditation
46,000 sq. ft., temperature controlled environment
45' by 90' vibration isolated laboratory

History of Laser Tracker

- 1979-80, Itek first demonstrated a 4-head trilateration laser tracking system for large optical surface measurement (Itek patented)
- In 1982, the single-beam laser tracker was conceived and demonstrated. NIST eventually patented a 3/5-D laser tracker (Kam Lau and Bob Hocken as co-inventors)
- In 1983-85, Chesapeake Laser (CLS) adopted the Itek concept and built a prototype with funding from the US Navy
- 1988/89, Boeing successfully conducted a full test and evaluation of the first API 3-D single-beam laser tracker
- 1989, first commercial SB laser trackers introduced by API/Leica, CLS followed

History Cont'd

- 1991 SMX acquired CLS
- 1994/5 Leica Tracker combined IFM+ADM capability, followed by SMX within a year
- 1999 API entered the market with the 2nd Generation Laser Tracker T2 (on-shaft mounting laser)
- 2002 API introduced T2+ with IFM+ADM capability, Faro acquired SMX
- 2004/5 Faro introduced X Series (fiber-guided laser)
- 2005 API introduced T3 and OT with Turbo ADM
- 2005/6 Leica and API introduced handheld probes
- 2008 Leica introduced Absolute Tracker

Revolution of an industry

"Since we adopted the use of laser tracker, we estimate a corporate saving of 4.5 billion dollars ..." quote from a senior manager at a major aircraft manufacturer in year 2000

The ease of use, accuracy and cost-effectiveness of laser tracker have totally changed the ways how aircrafts are built. Other industries also experience the same magnitude of economic and productivity impacts. After 20 years of its introduction, the impact still continues ...

Theory of Laser Tracking System

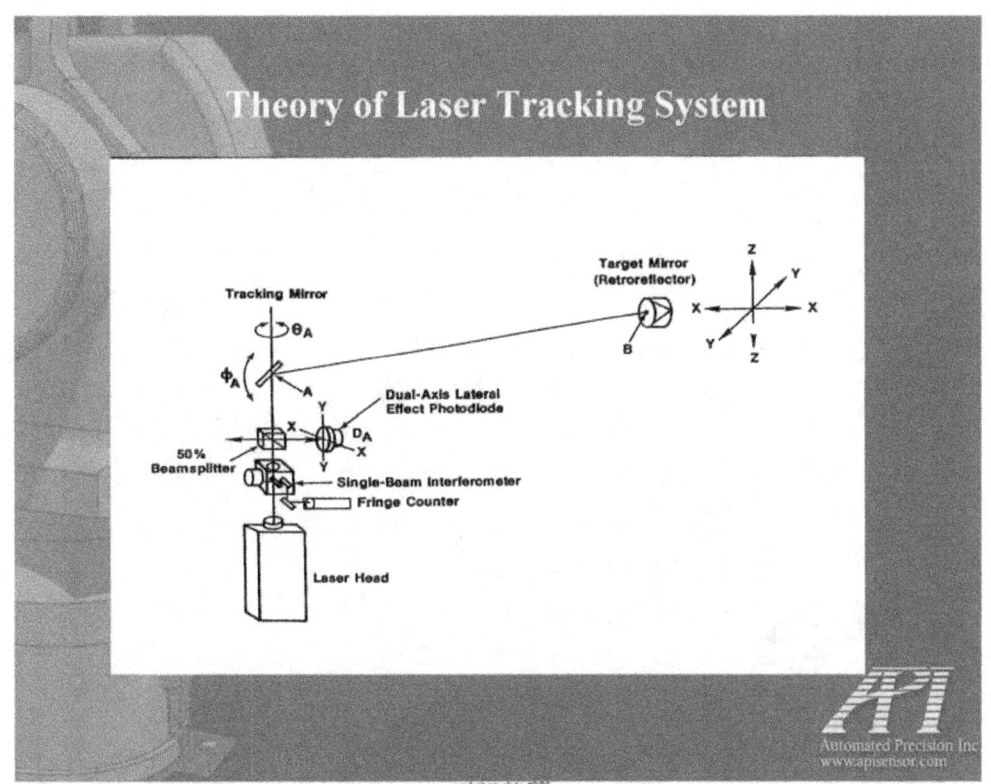

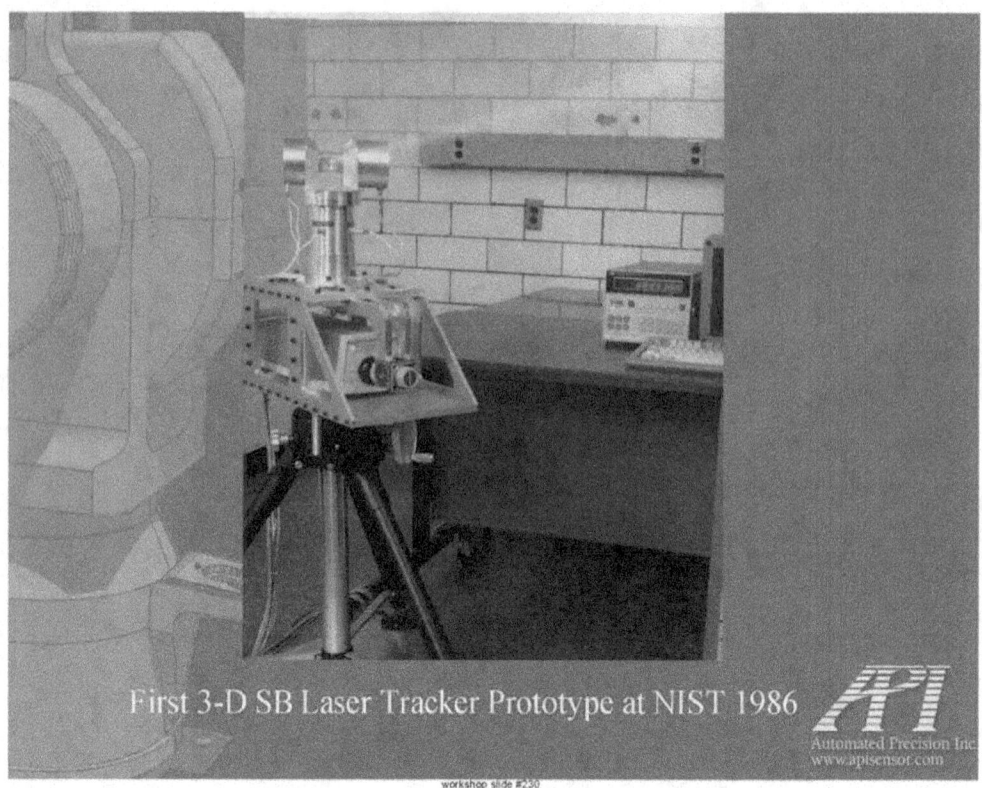

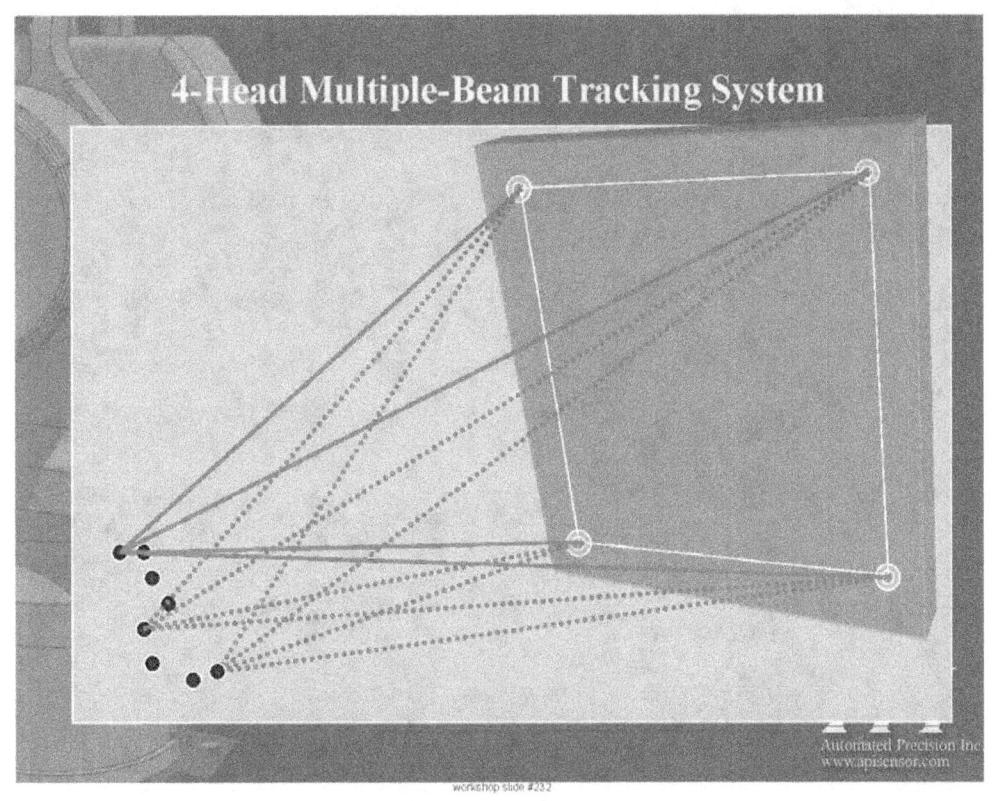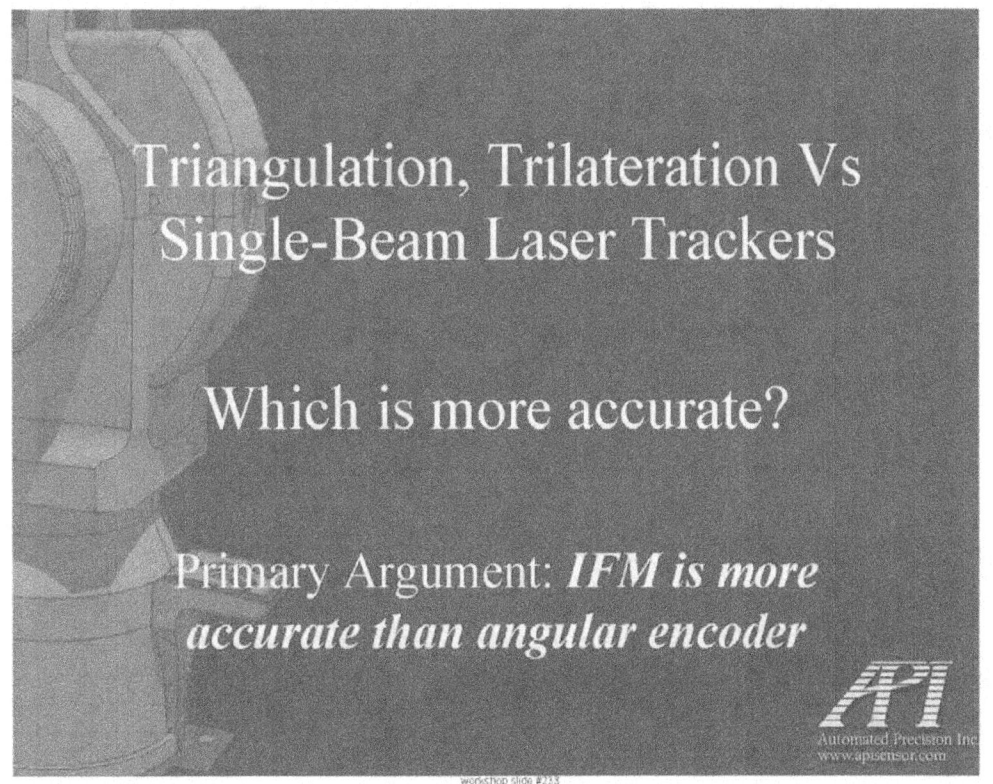

Assumptions for Discussion

- No environmental effects
- No consideration of metrology frame inaccuracy or instability
- No consideration of artifacts or setup errors

- Angular uncertainty – 1 arc-second
- Linear uncertainty – 1 ppm

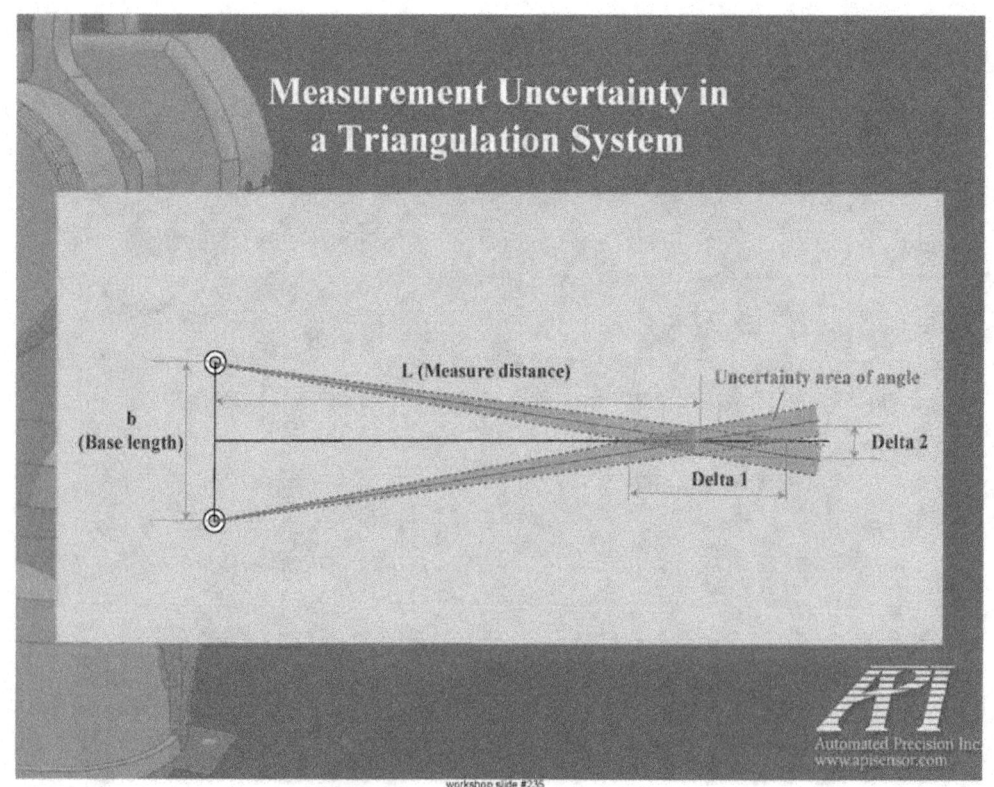

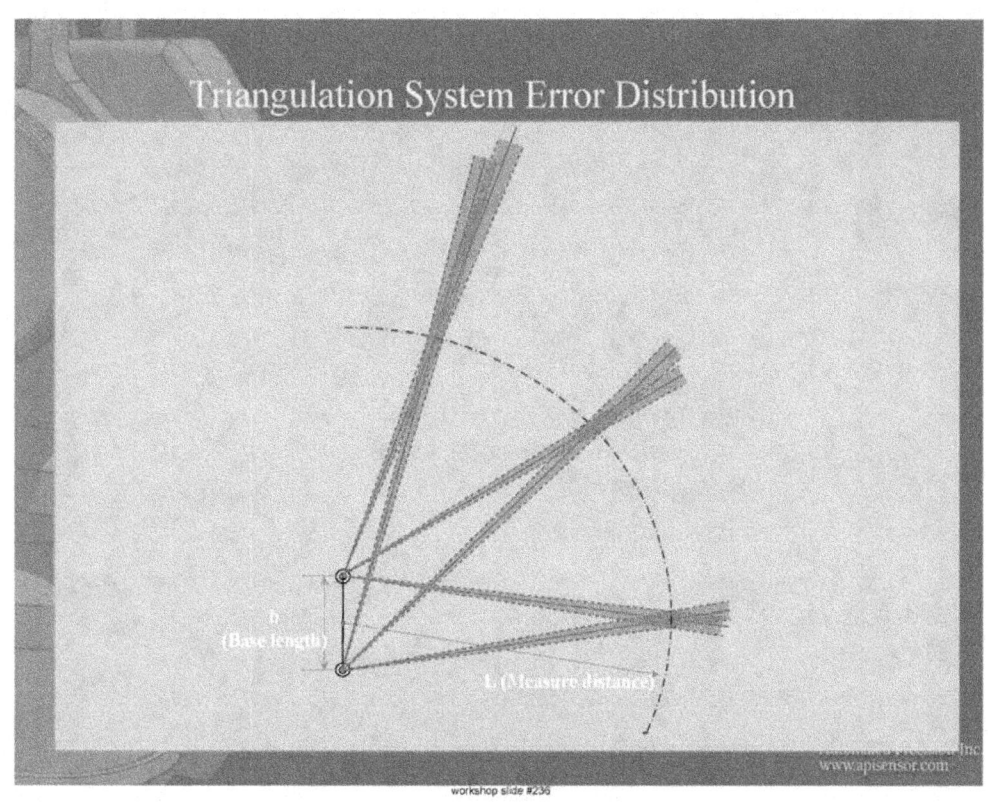
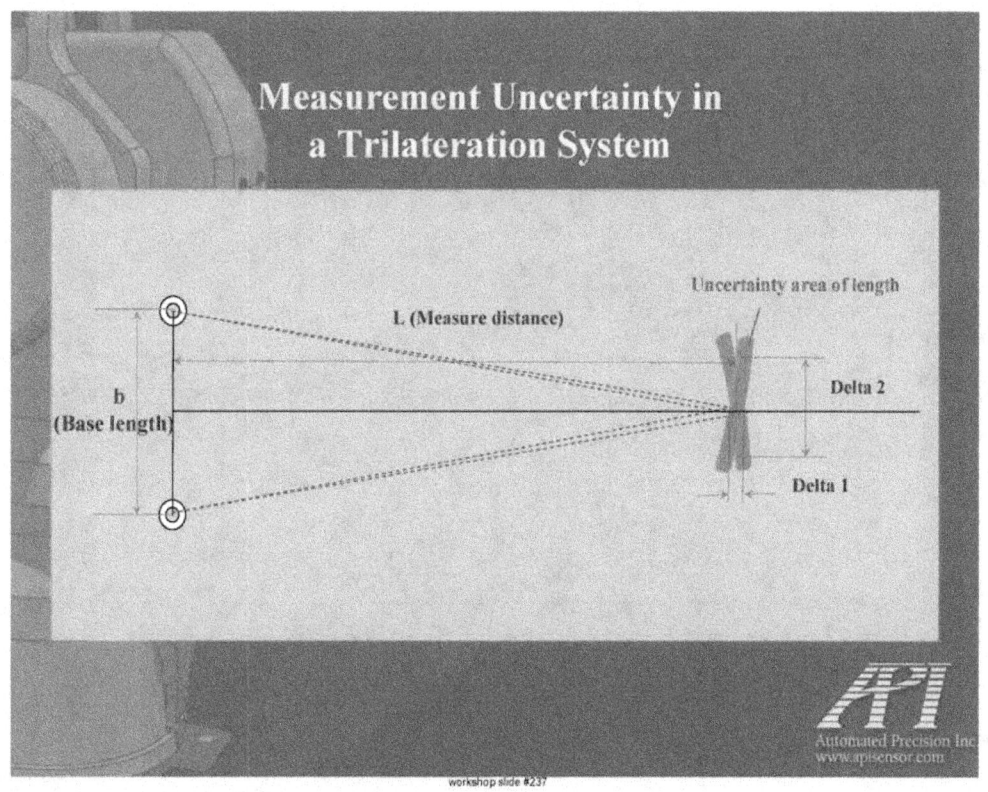

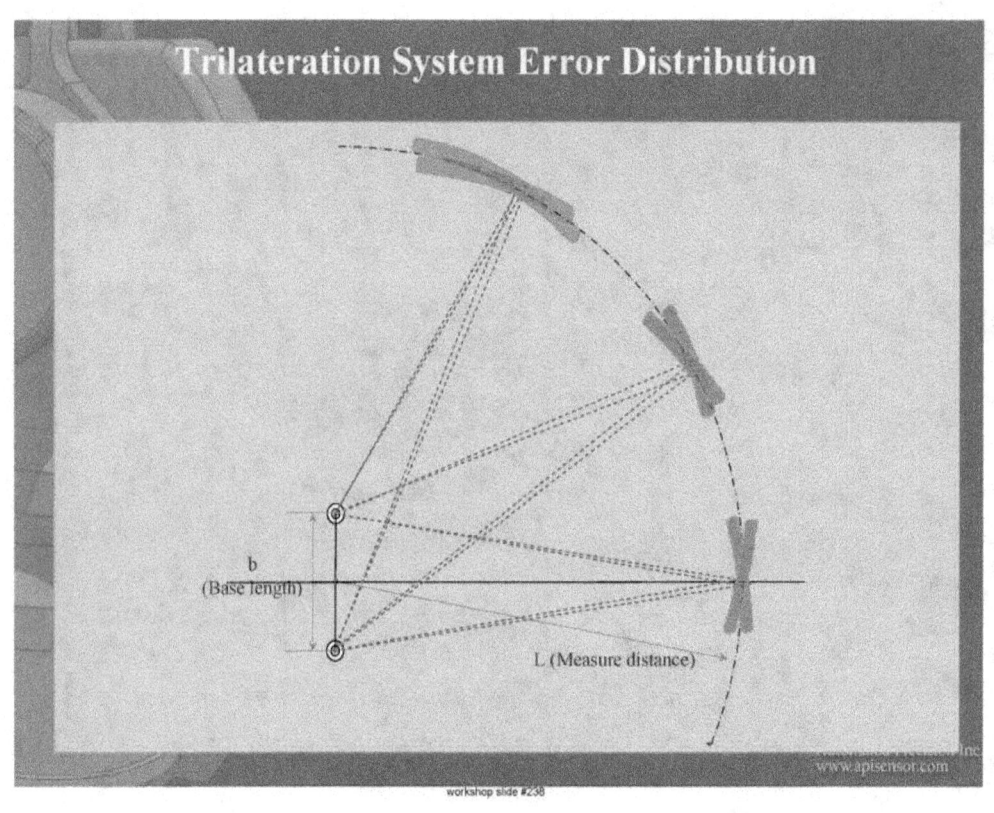
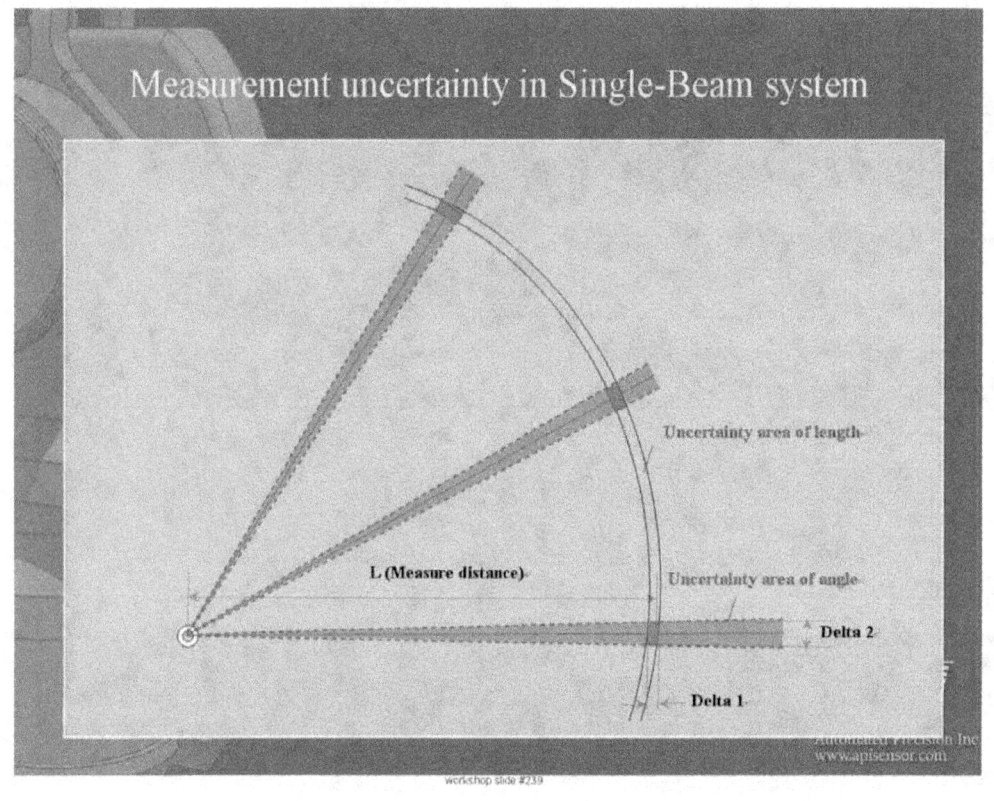

Theoretical accuracy comparison of SB and MB tracking system

Measure Distance (m)	Triangulation system b=5m Delta 1 (mm)	Trilateration system b=3m Delta 2 (mm)	Single-Beam system Delta 2 (mm)
1.5	0.0330	0.0030	0.0145
2	0.0398	0.0042	0.0194
2.5	0.0485	0.0057	0.0242
3	0.0591	0.0075	0.0291
3.5	0.0718	0.0097	0.0339
4	0.0863	0.0122	0.0388
4.5	0.1028	0.0150	0.0436
5	0.1212	0.0182	0.0485
10	0.4121	0.0682	0.0970
20	1.5756	0.2682	0.1939
30	3.5148	0.6015	0.2909
40	6.2296	1.0682	0.3879

Theoretical Accuracy Comparison of SB and MB Tracking Systems

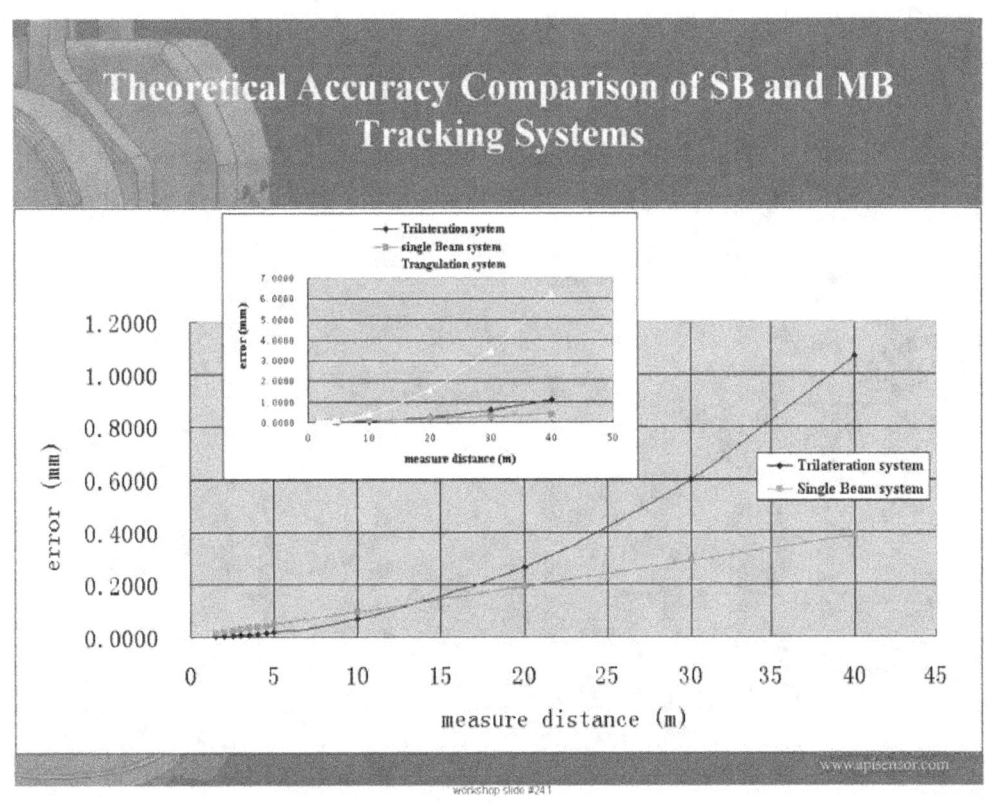

Pros and Cons on SB Vs MB for Large-Scale Metrology

- MB dominated by non-linear regions, optimum accuracy at 60° envelop; SB is more linear
- MB requires artifact calibration to define base distances therefore reducing accuracy, SB does not
- Uncertainties in MBs crossing apex of SMR, metrology base frame stability, etc. compromise overall accuracy.
- Portability, cost, ease of use and accuracy is field certifiable make SB tracker more favorable for industrial applications

Evolutions of Tracker Head Designs and Heat Management

- Remote Vs On-shaft laser mounted
- Key principles to better head design
    - Axis symmetry for thermal stability
    - Shortest optical deadpath, minimum moving mirrors
    - Abbe' Principle compliance
    - Structural rigidity but no mass
- Heat Management
    - Remove (impossible) or minimize heat source
    - Incorporate heat source into the design

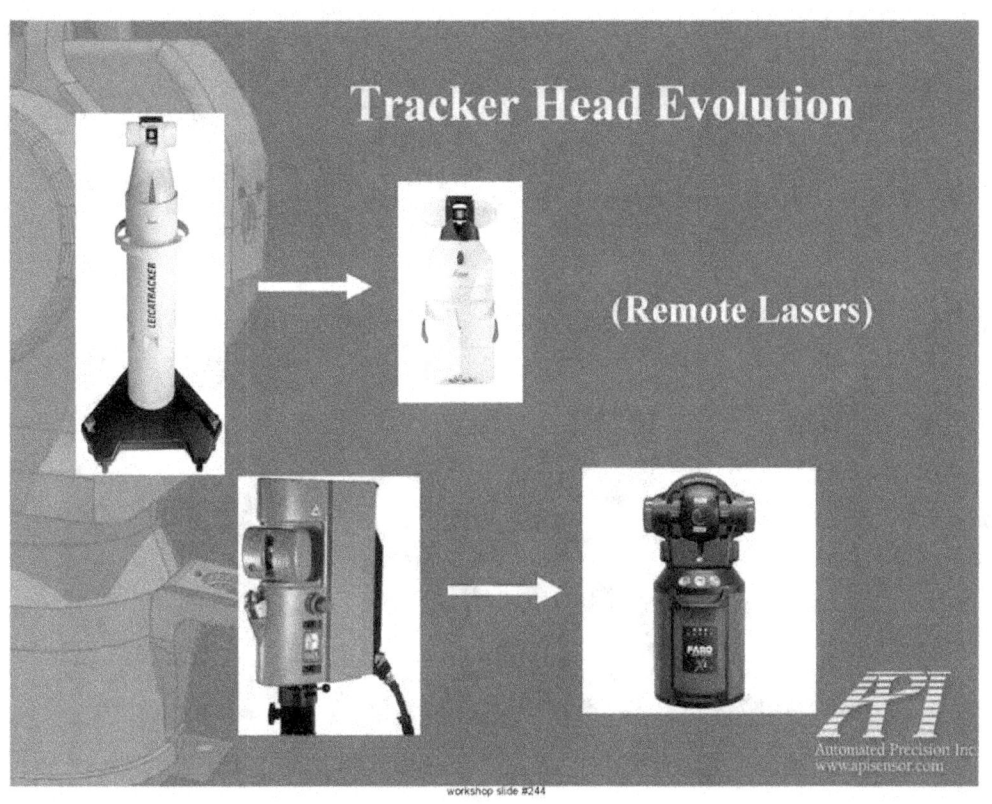
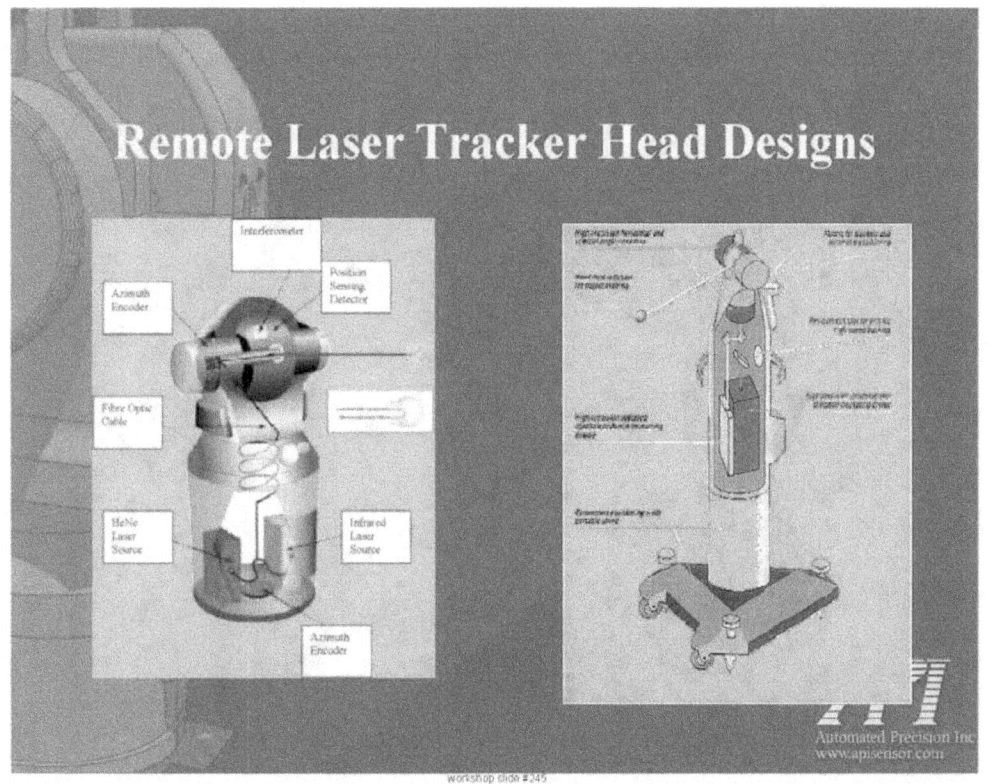

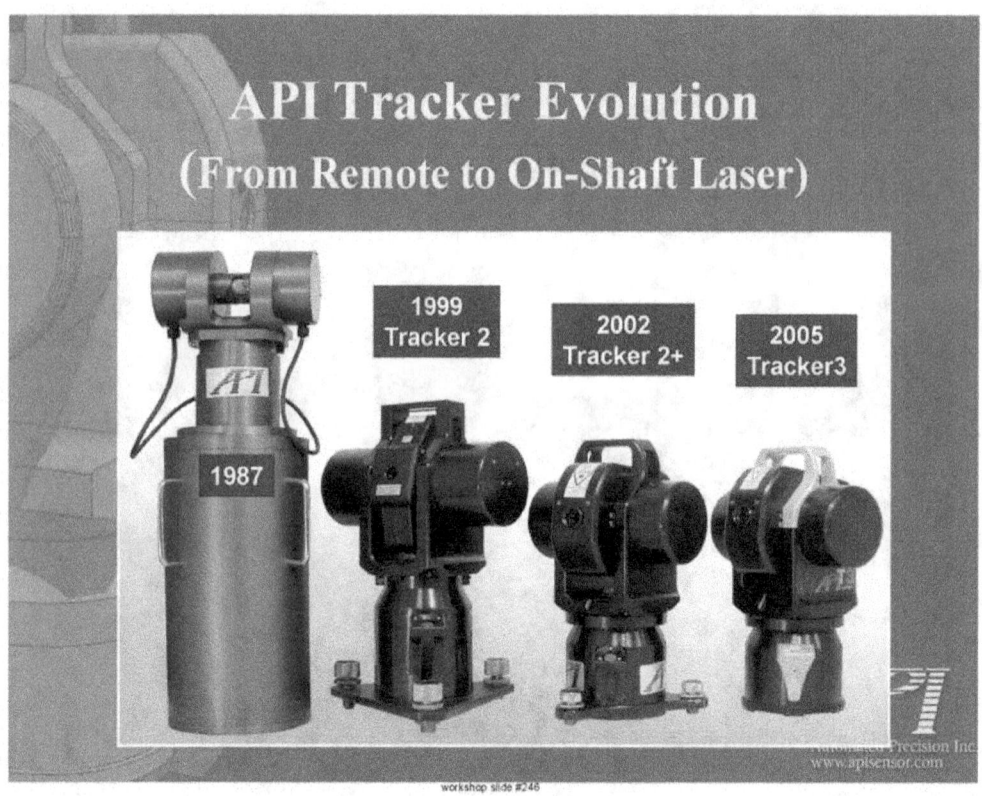

Challenges and Benefits
Remote Vs On-Shaft Mounting

- **Challenges**
  - Complexity of overall design (optic, ME, EE)
  - Much demanding digital servo system
  - Stringent heat management scheme required
- **Benefits**
  - Compact, light weight
  - Larger EL angle operation than reflective mirror
  - 1:1 encoder and beam relationship
  - Fewer error sources resulted in better long stability
  - Minimum Abbe' Errors

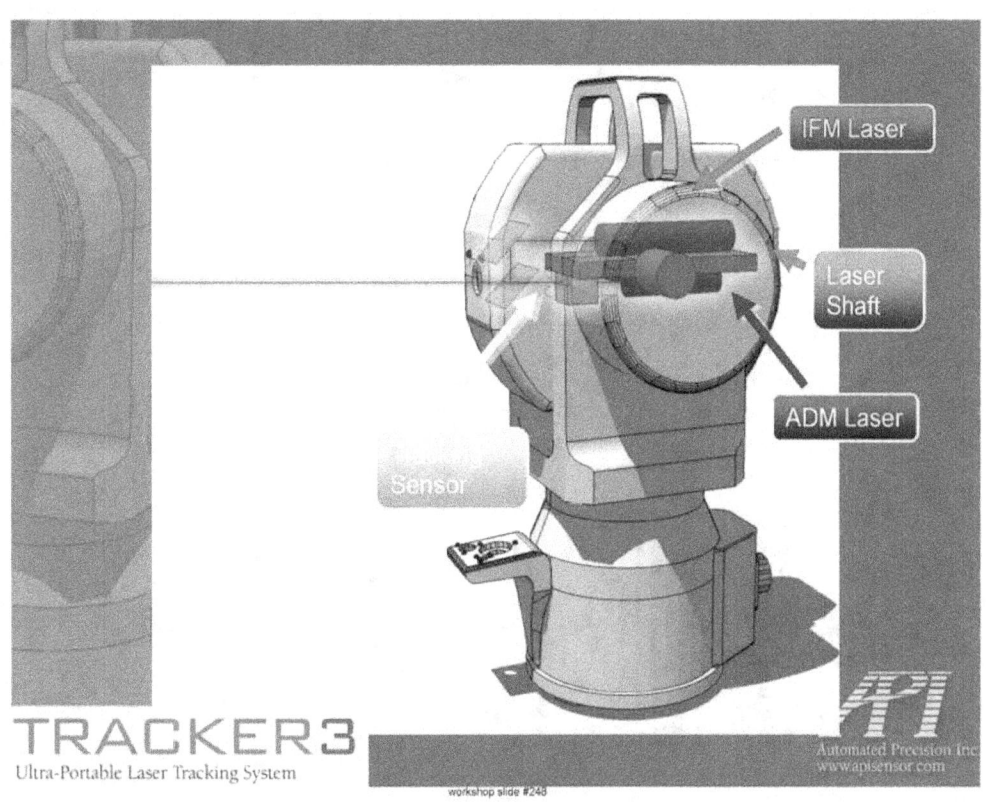

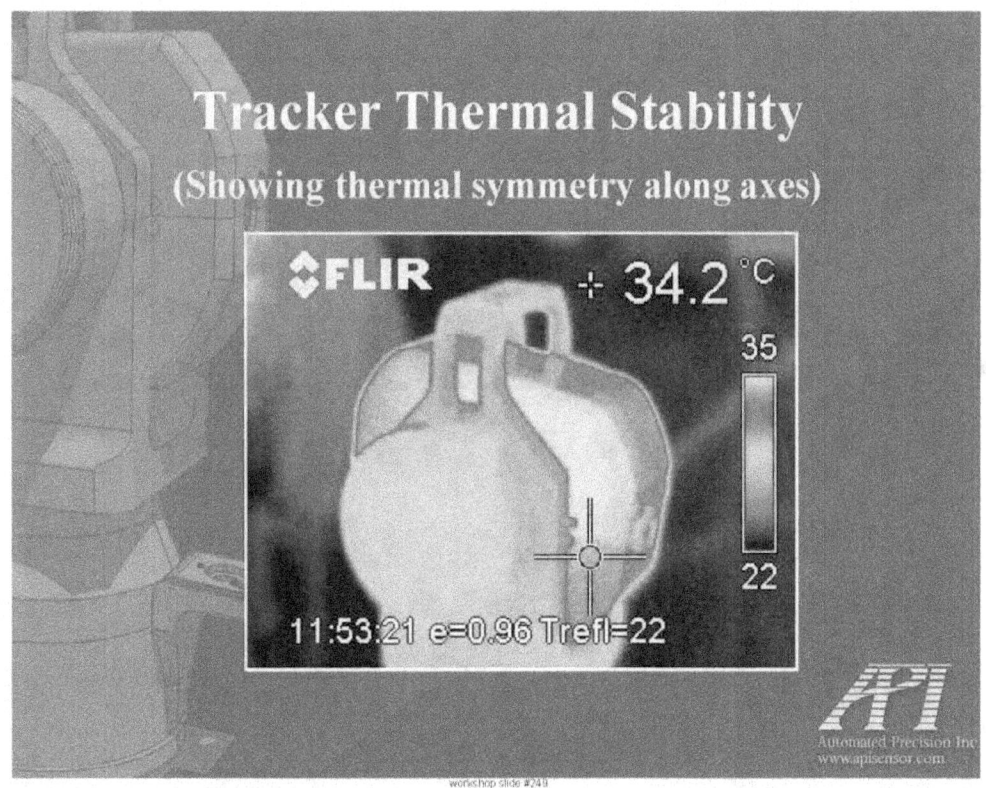

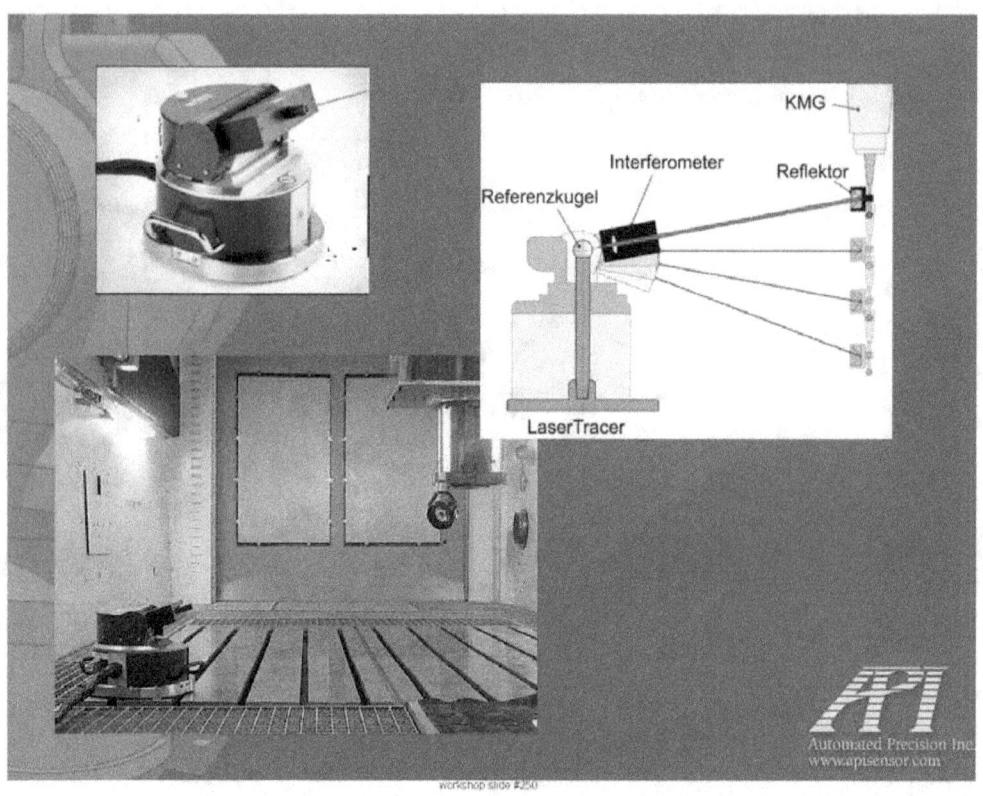

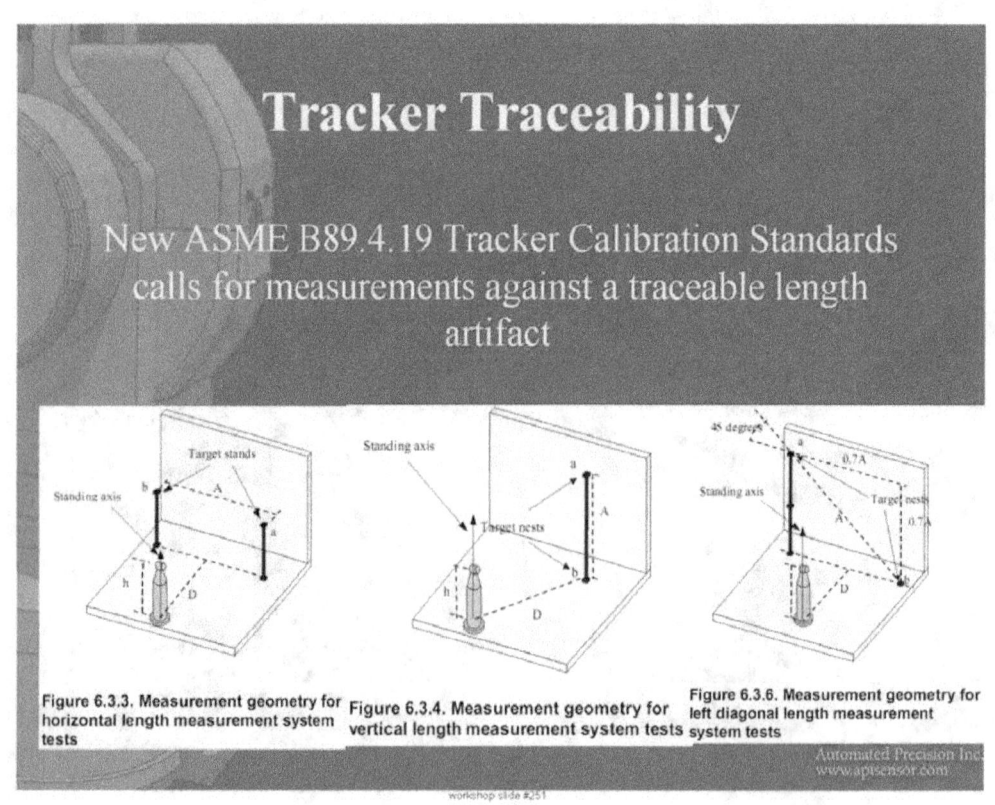

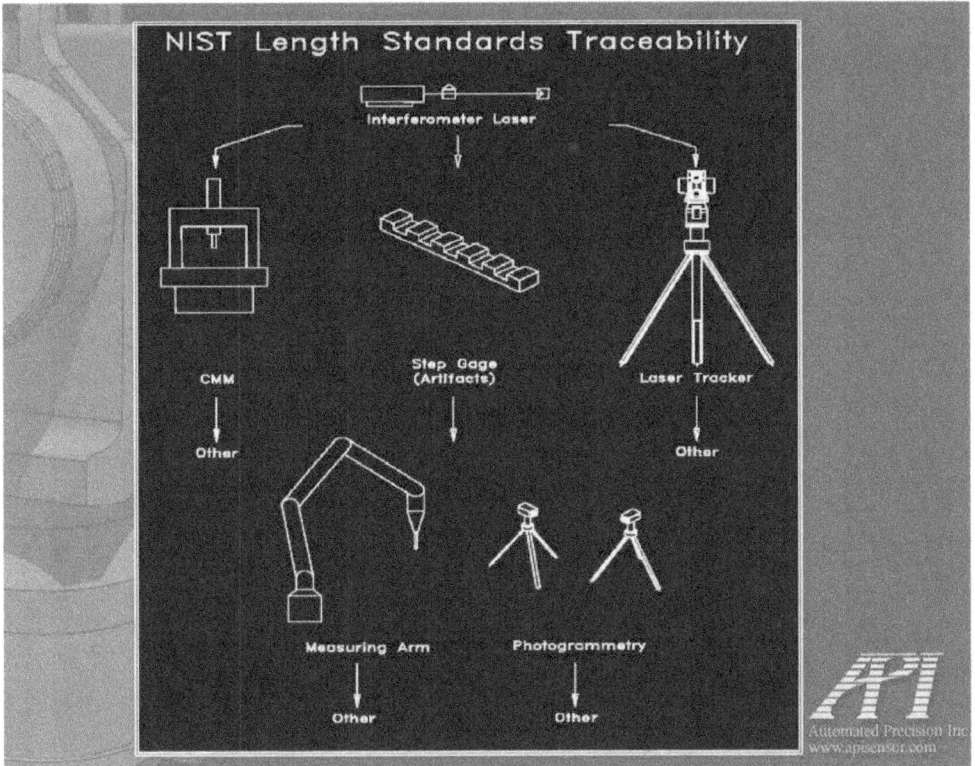

USMN Vs ASME B89

- Laser interferometer is the internationally accepted length standard
- Laser tracker is the only portable 3-D measuring instrument with built-in length standard
- **One of the major reasons for the popularity of laser tracker !!**
- USMN (Unified Spatial Metrology Network) utilizes the embedded interferometer for tracker self-certification
- Caution: ADM-only tracker is an exception

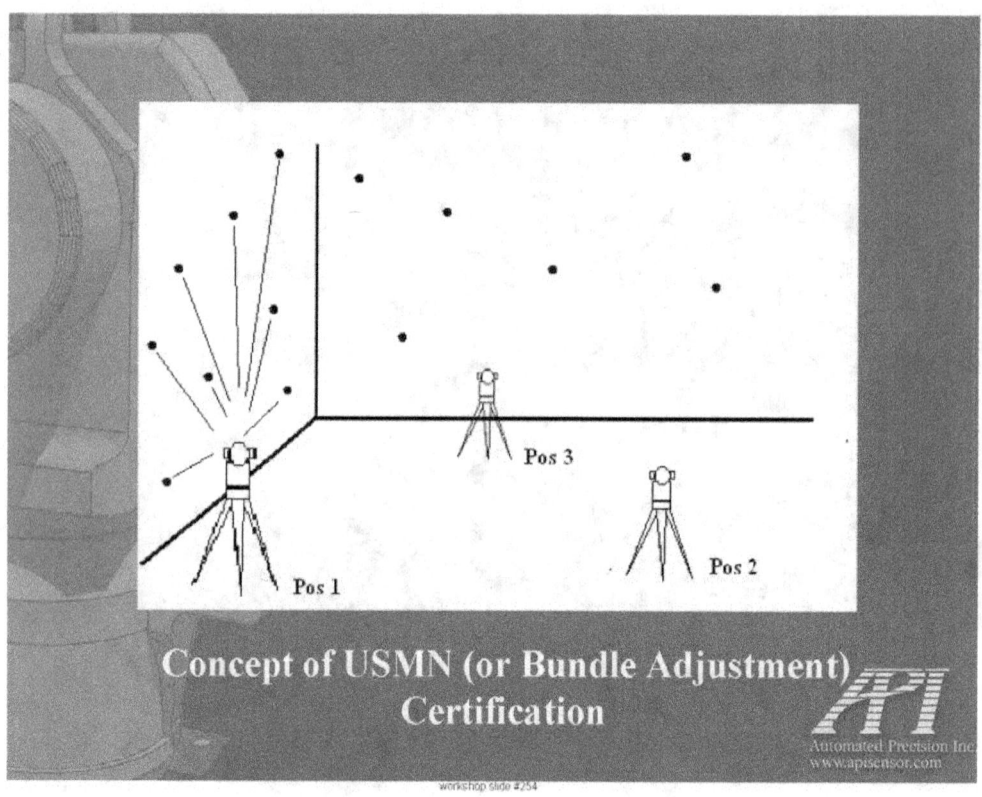

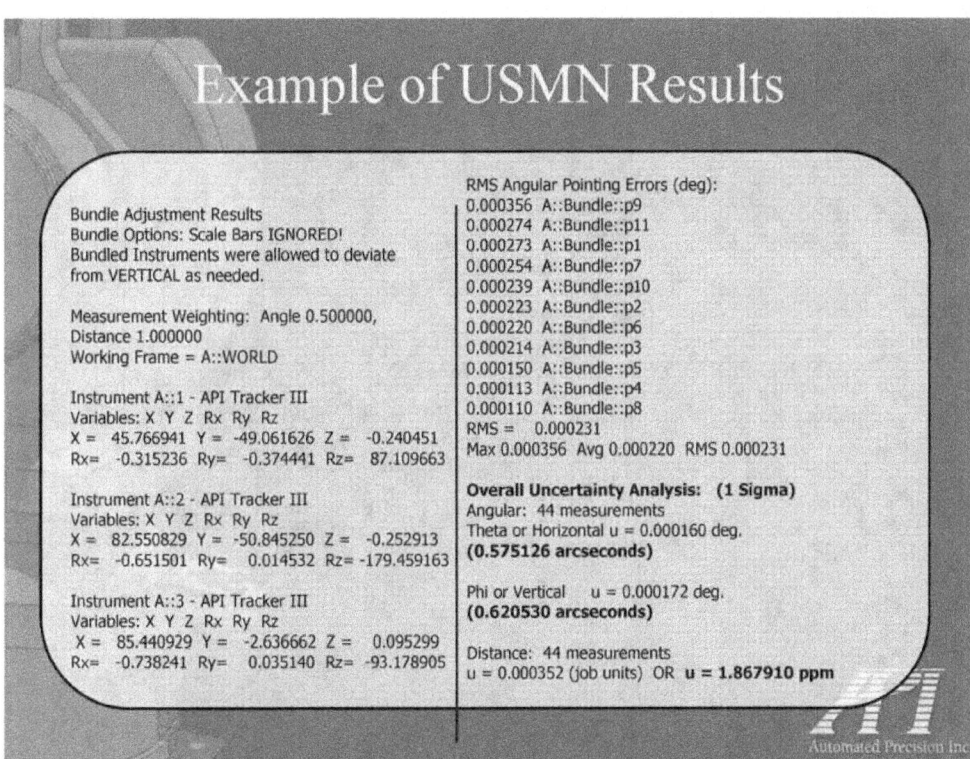

workshop slide #258

workshop slide #259

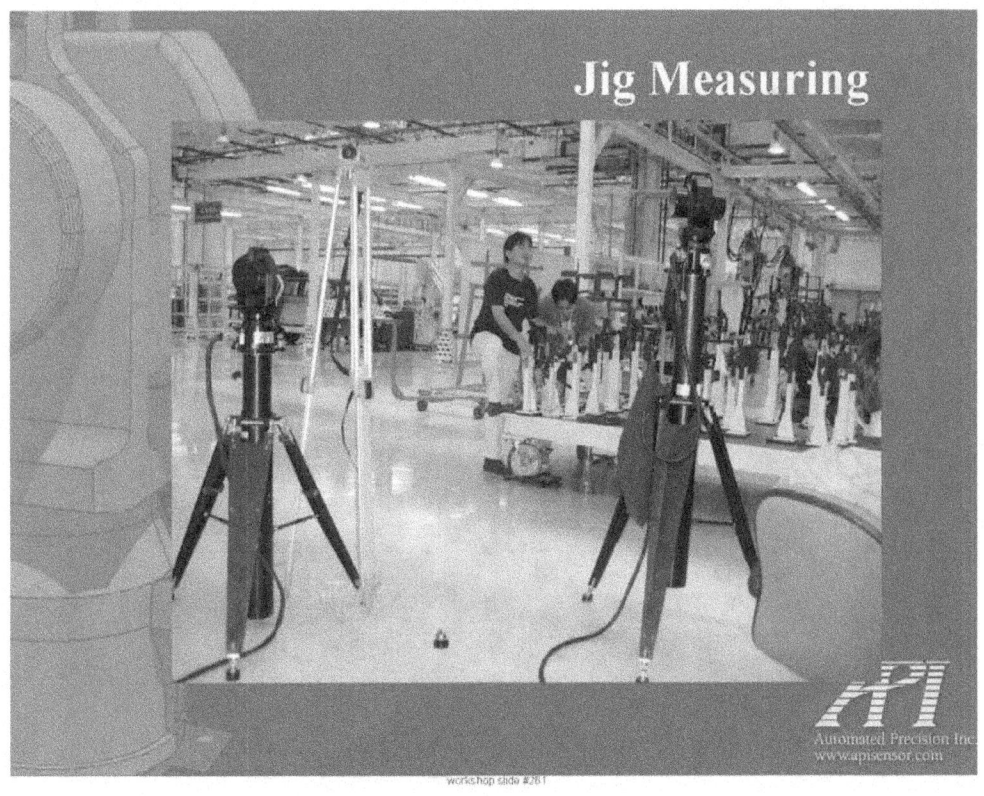

workshop slide #264

workshop slide #265

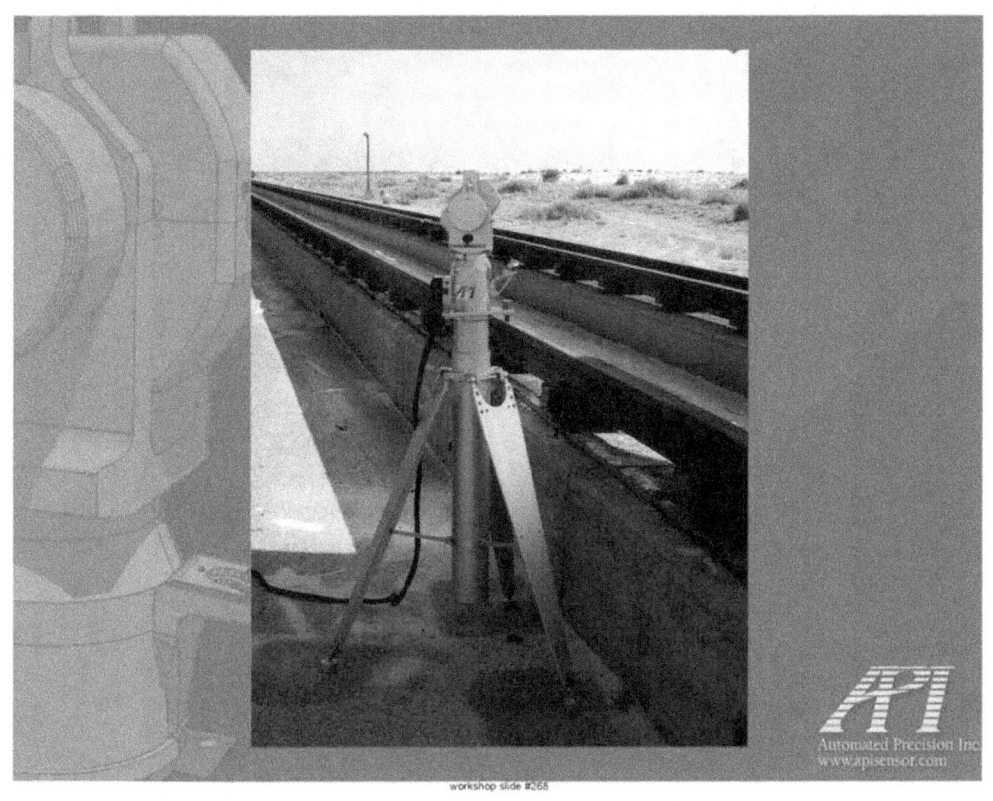

ABB ROBOT Calibration

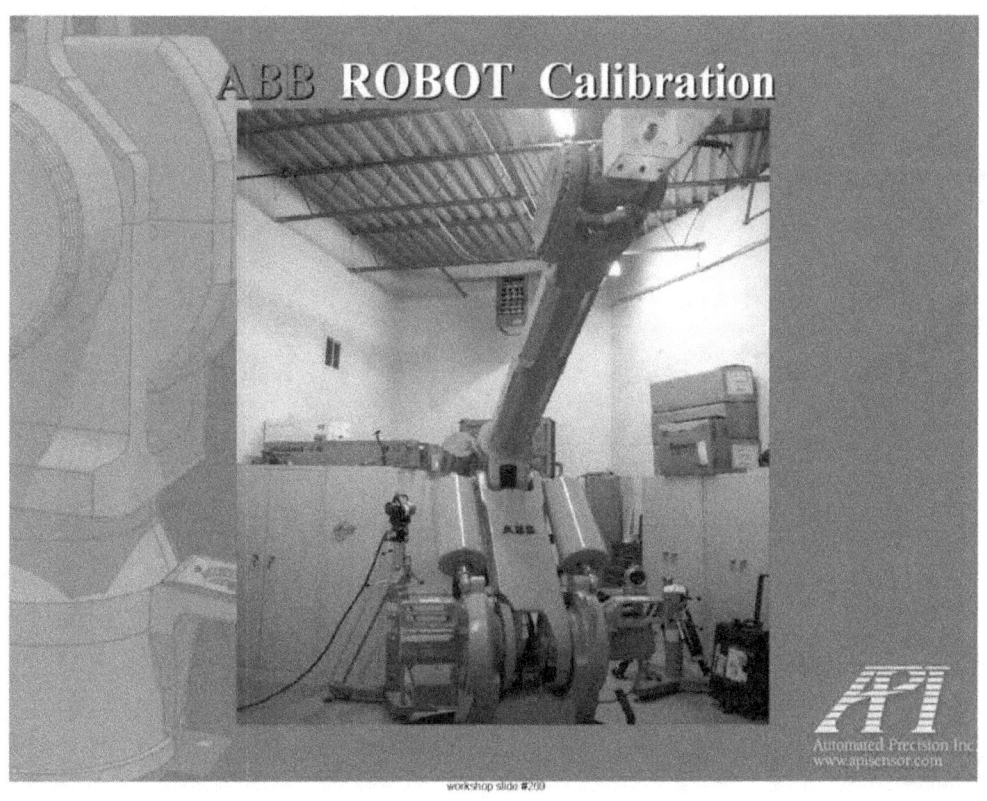

New Development in Machine Tool Error Mapping with Laser Tracker

- Traditional wisdom – tracker doesn't have enough precision
- Work at Boeing St. Louis successfully demonstrates it is possible
- Phil Freeman and Sam Easley successfully mapped 11 shop floor machines in various Boeing plants providing 3 to 4 times improvements of volumetric accuracy

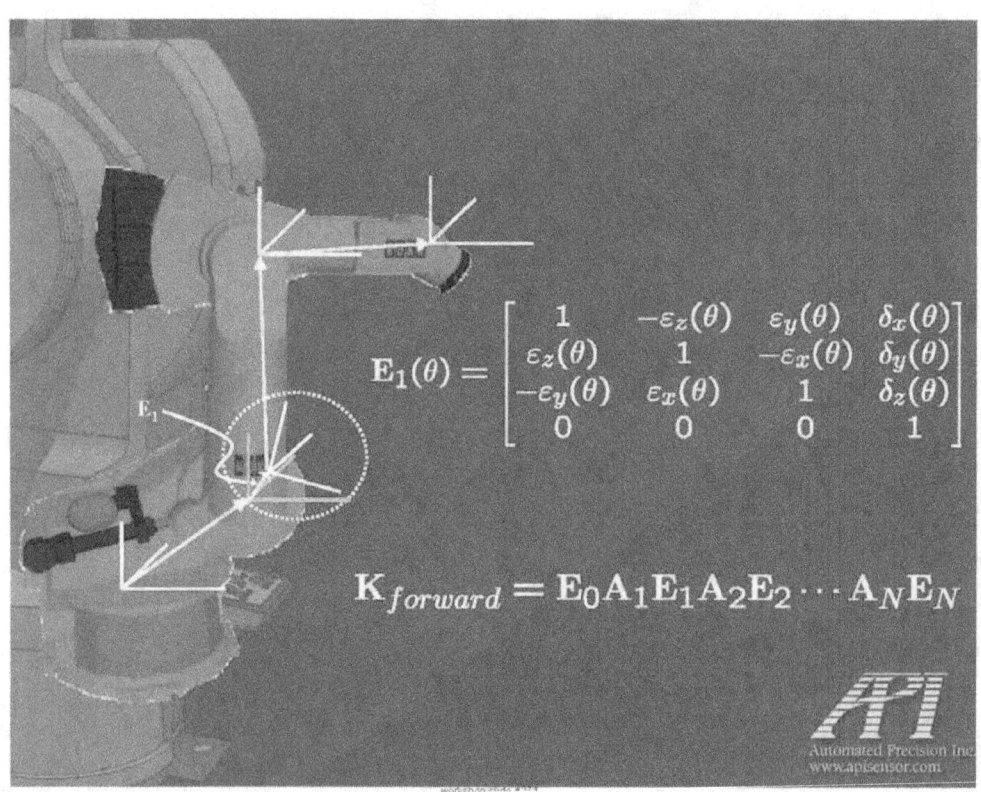

$$E_1(\theta) = \begin{bmatrix} 1 & -\varepsilon_z(\theta) & \varepsilon_y(\theta) & \delta_x(\theta) \\ \varepsilon_z(\theta) & 1 & -\varepsilon_x(\theta) & \delta_y(\theta) \\ -\varepsilon_y(\theta) & \varepsilon_x(\theta) & 1 & \delta_z(\theta) \\ 0 & 0 & 0 & 1 \end{bmatrix}$$

$$K_{forward} = E_0 A_1 E_1 A_2 E_2 \cdots A_N E_N$$

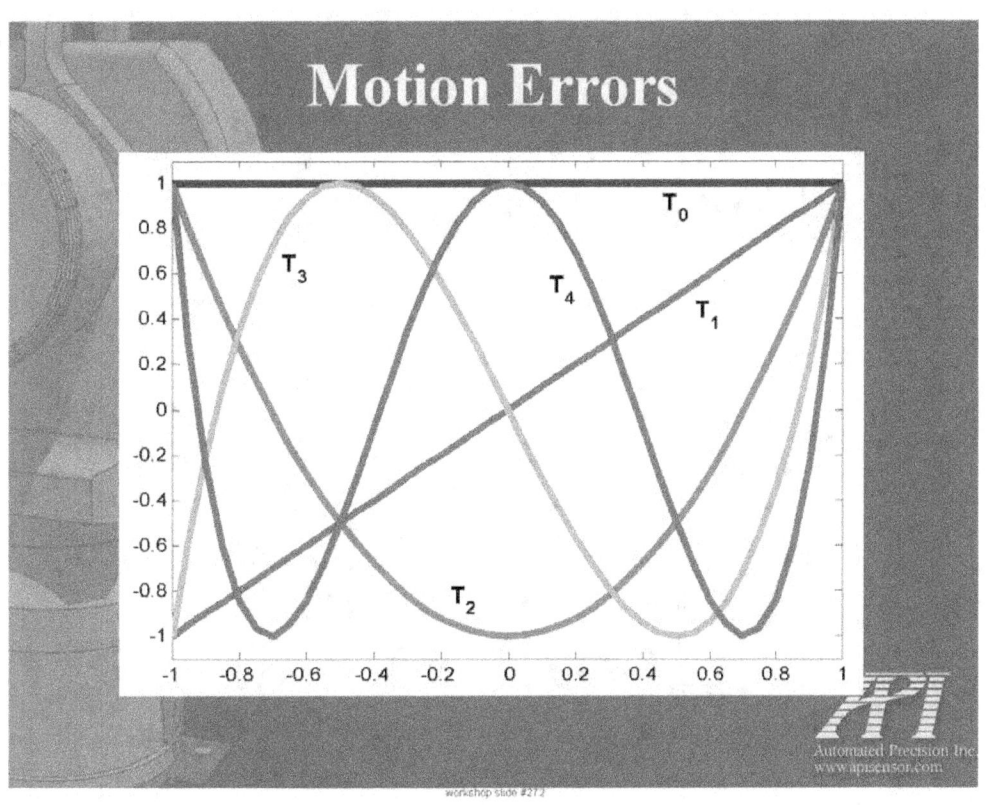

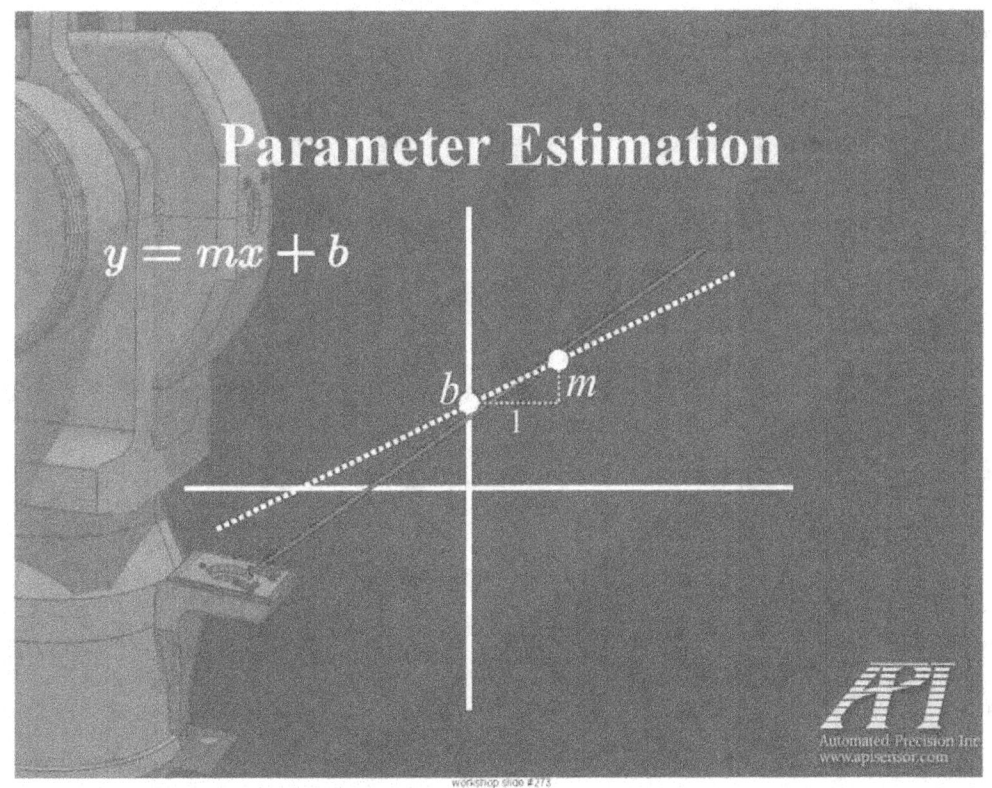

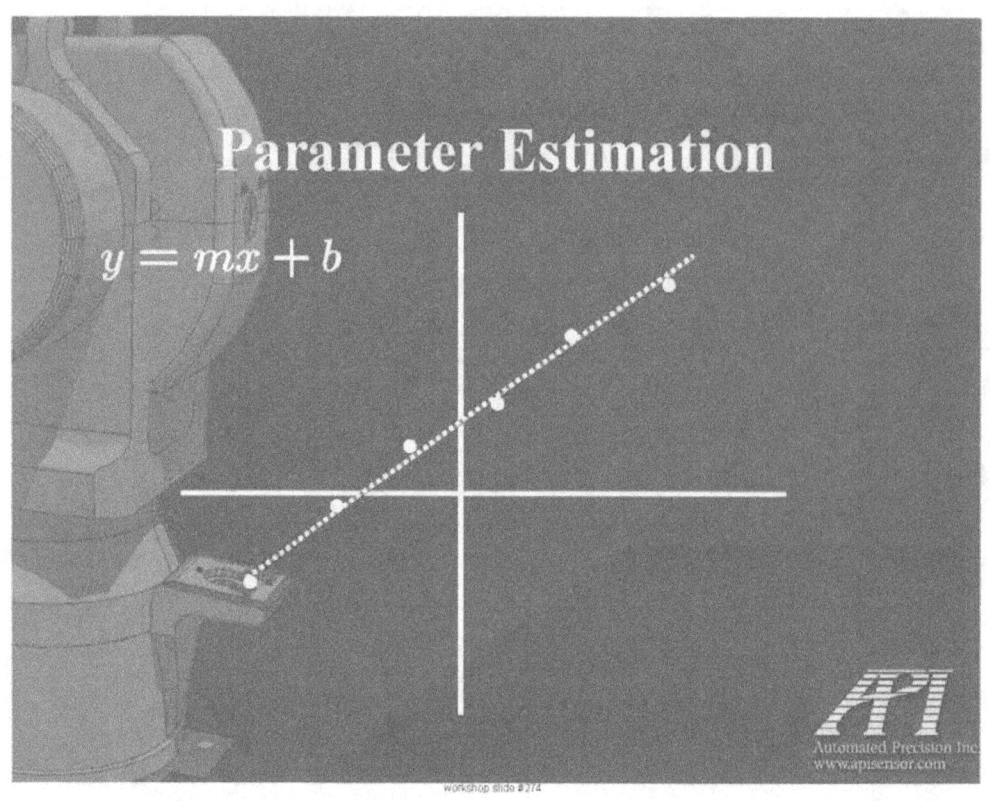

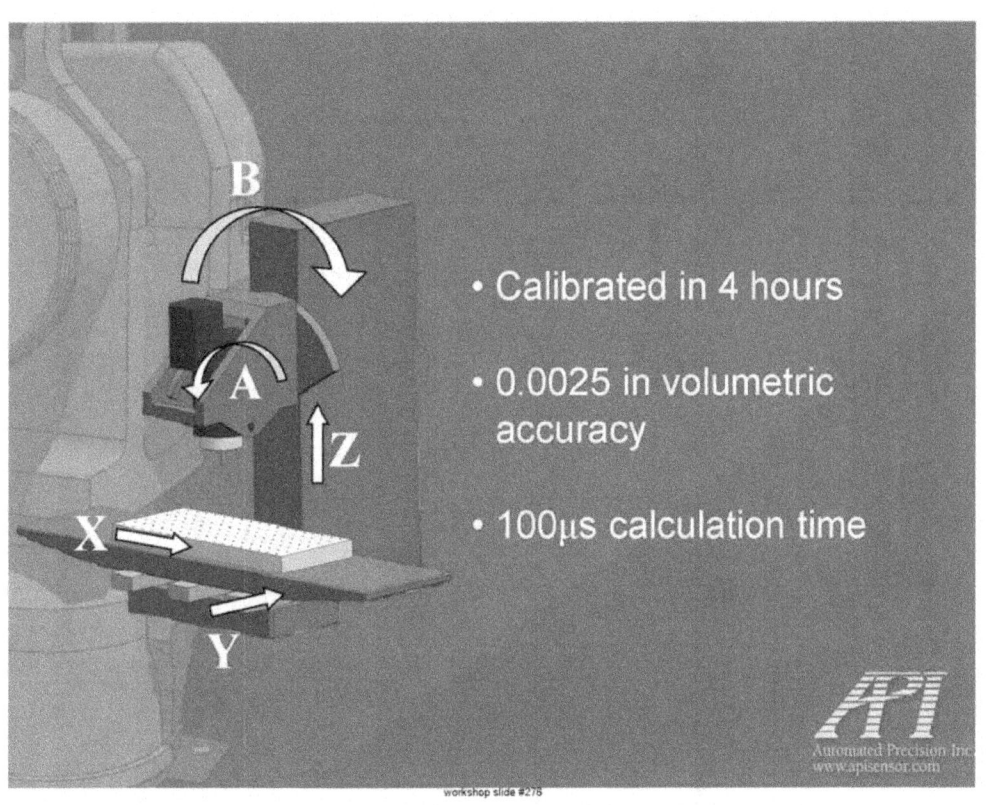

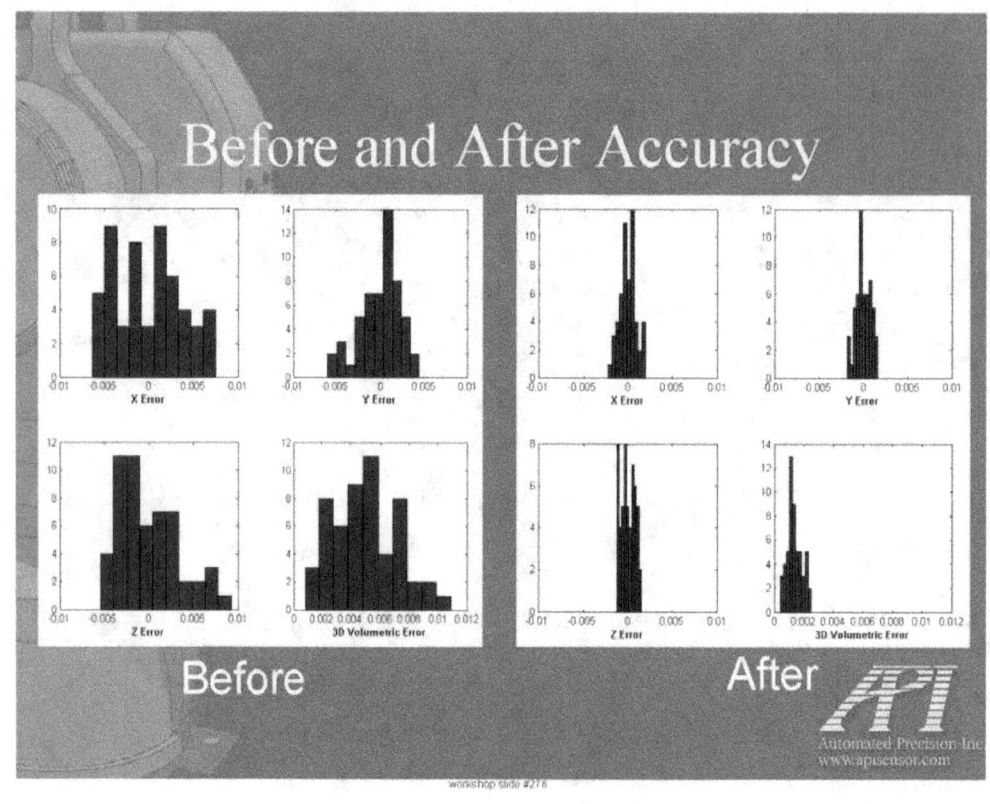

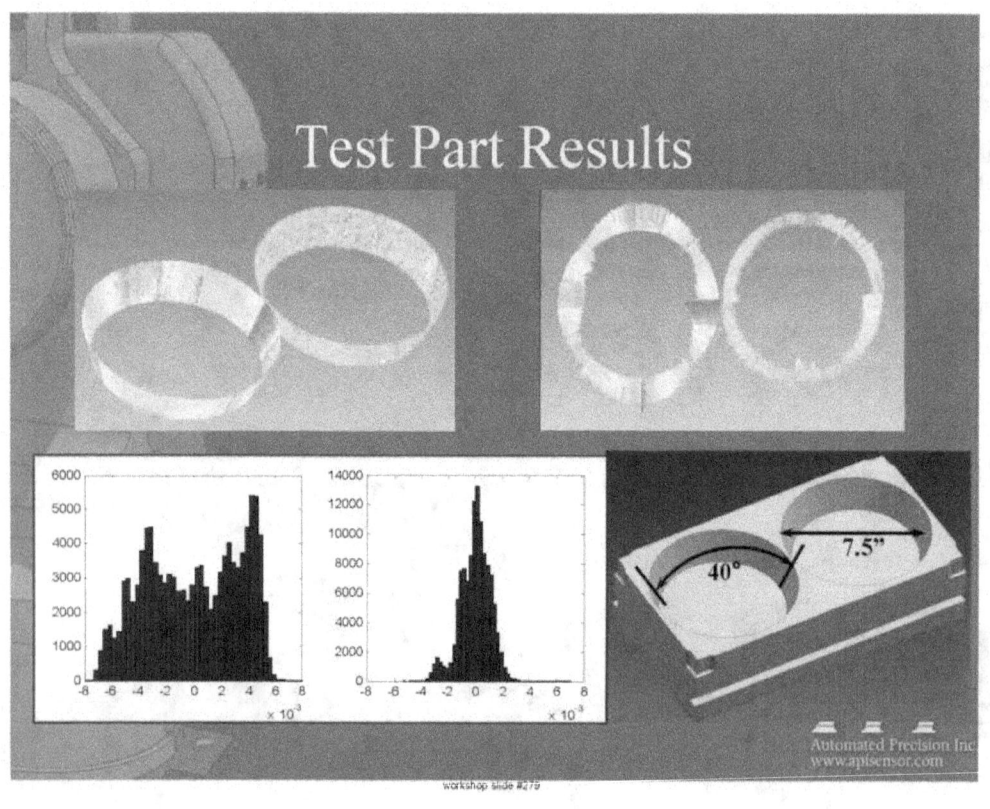

Advanced Tracker Accessories

Advanced accessories give tracker more versatilities in dealing with difference measuring challenges
- Hidden points
- Surface scanning
- Non-contact
- High data-rate point-cloud
- 6-D measurements
- Programmable automation

SMARTTRACK
6D Laser Tracking Sensor

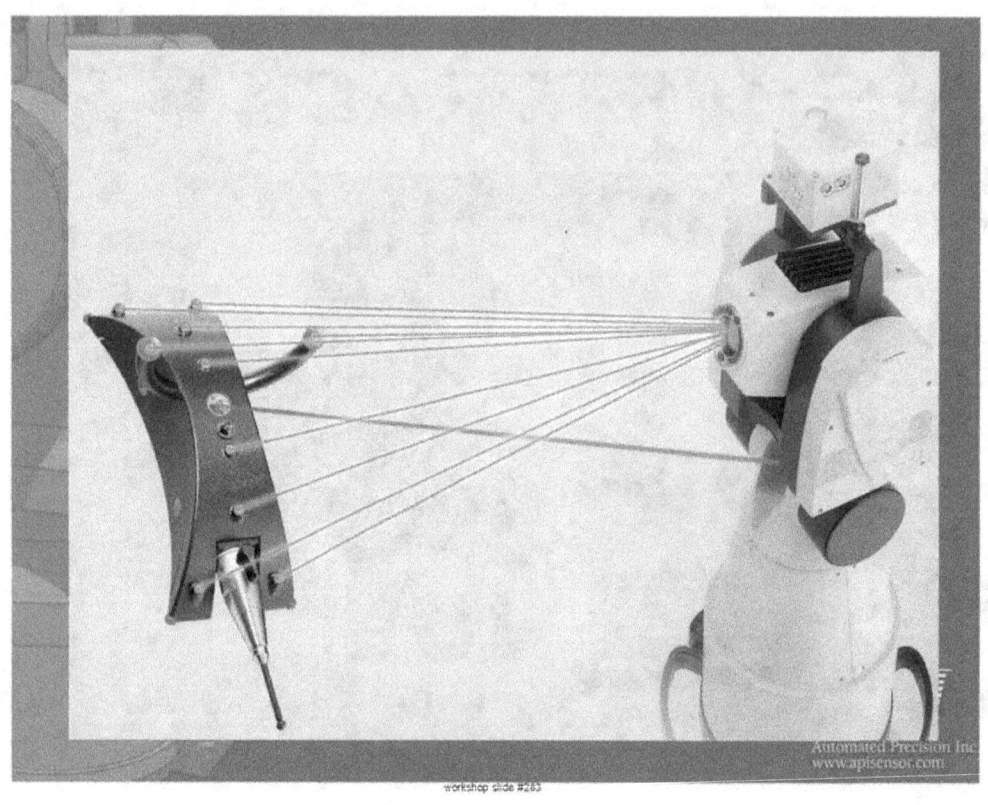

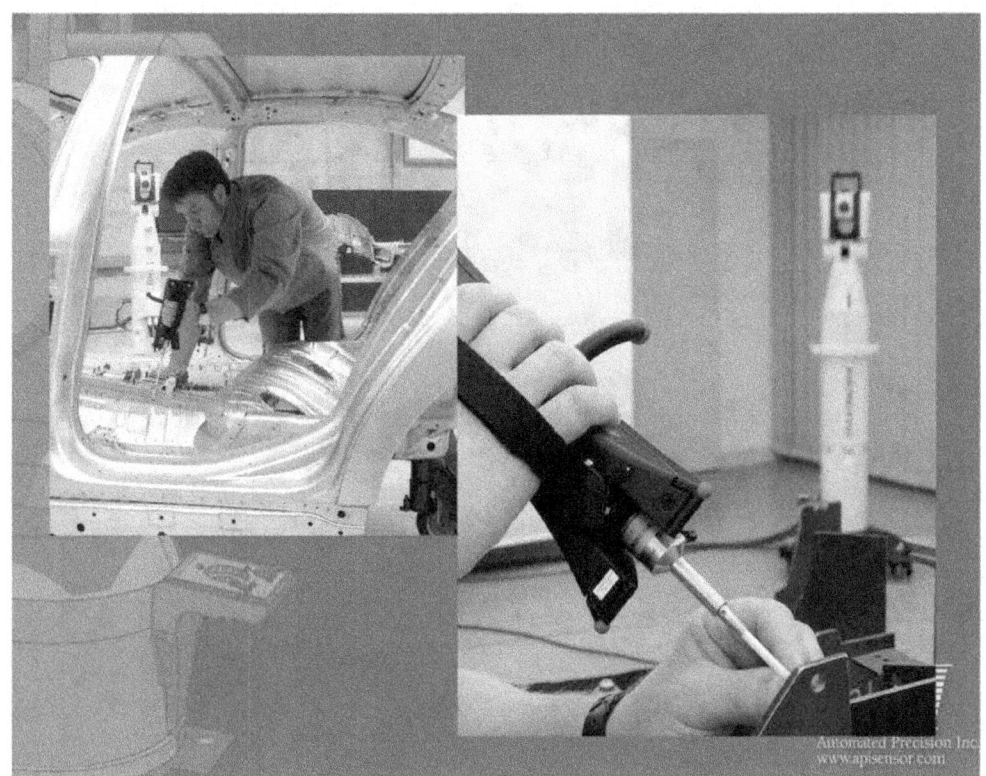

workshop slide #284

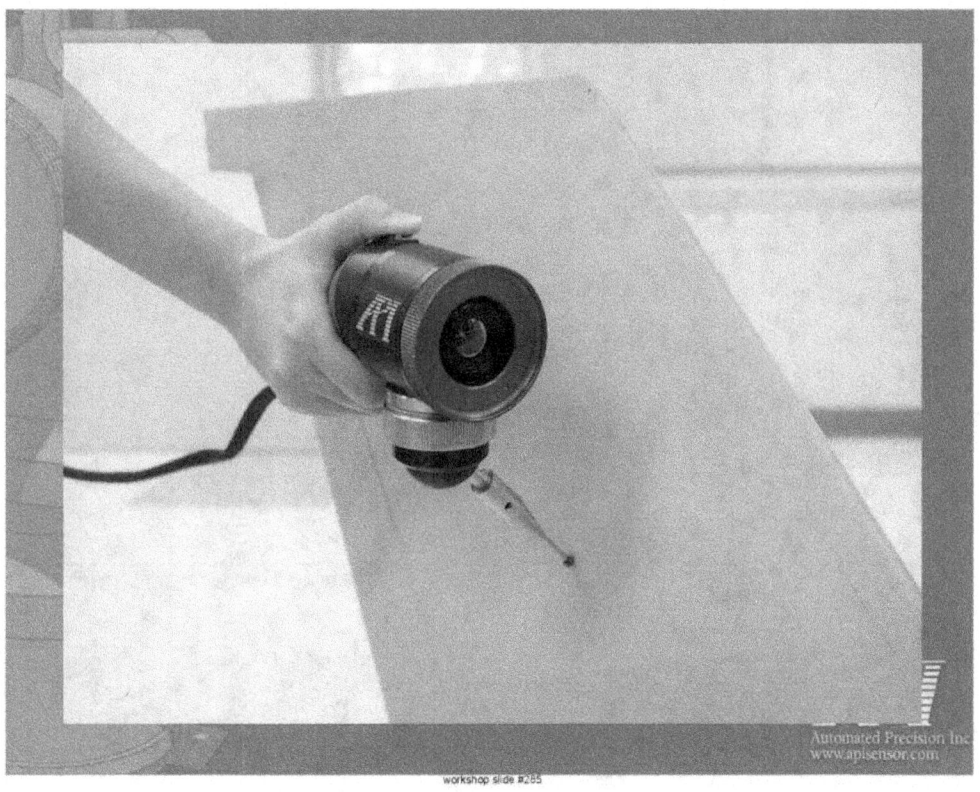

workshop slide #285

workshop slide #286

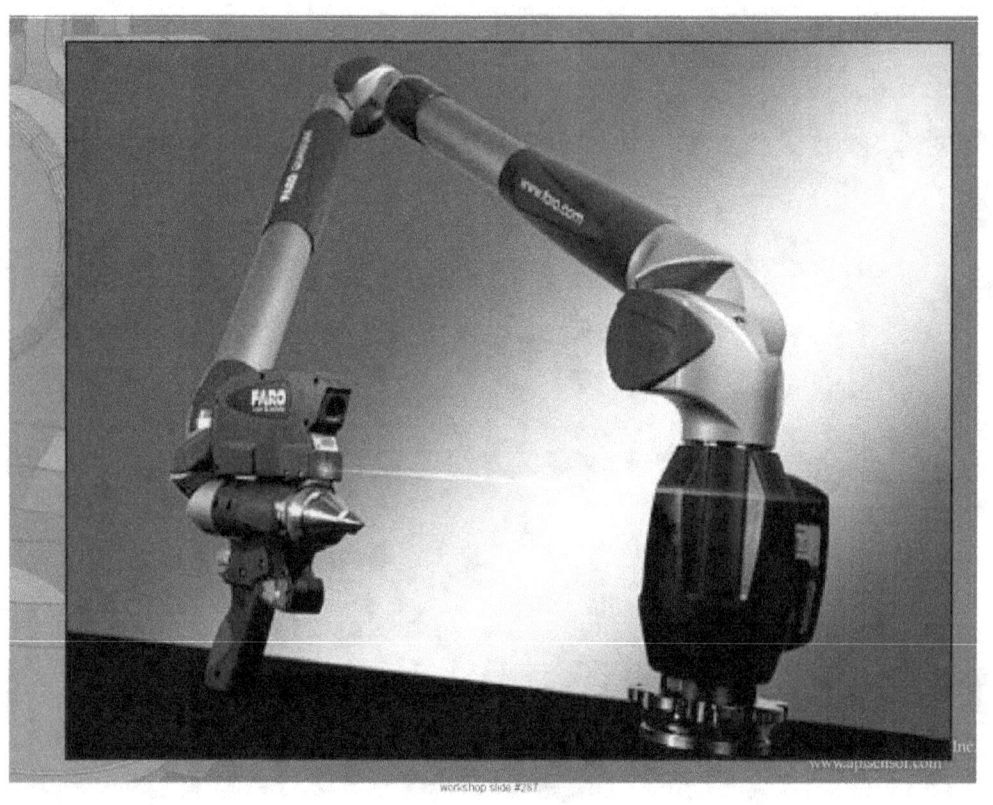
workshop slide #287

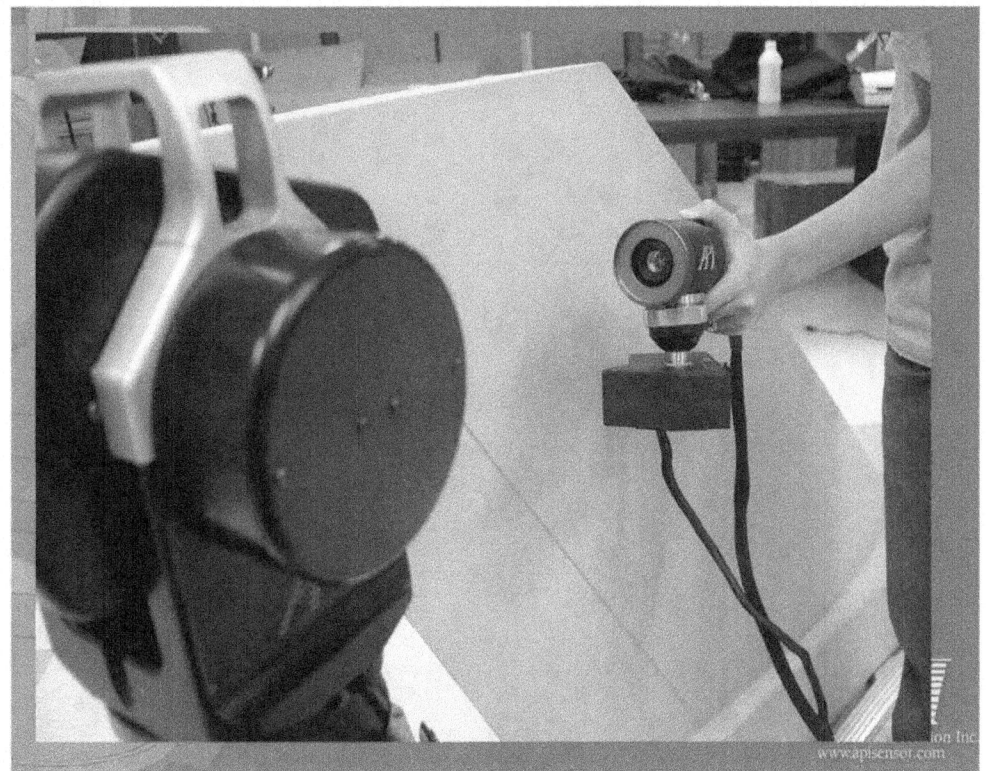

Trends of Laser Tracker Technology and Applications

- Applications from aerospace to automobiles, antenna, shipyard, machine tools, heavy industries, ...
- Smaller, lighter, lower cost, more precise, longer range, field certifiable, more features
- Advanced accessories like hand-held probes (contact or non-contact), multiple-degree of freedom tracking, integration with arms, cameras, electronic levels, photogrammetry, optical surface scanners
- New market continues to grow as applications expand
- Becomes a major threat to CMM and theodolite markets

www.ingramcontent.com/pod-product-compliance
Lightning Source LLC
Chambersburg PA
CBHW081612200526
45167CB00019B/2434